Effective Portfolio Management Systems

The **LITTLE BIG BOOK** Series

Other books in THE **LITTLE BIG BOOK** SERIES

Performance Acceleration Management, H. James Harrington (2013)

Closing the Communication Gap, H. James Harrington and Robert Lewis (2013)

Maximizing Value Propositions to Increase Project Success Rates,
H. James Harrington and Brett Trusko (2014)

Making the Case for Change: Using Effective Business Cases to Minimize Project and Innovation Failures, H. James Harrington and Frank Voehl (2014)

Techniques and Sample Outputs that Drive Business Excellence,
H. James Harrington and Charles Mignosa (2015)

Recreating Information Technology: As a Catalyst to Deliver Total Organizational Change, H. James Harrington, Richard Harrington Jr., and Ron Skeddle

Creativity, Innovation and Entrepreneurship: The Only Way to Renew Your Organization, H. James Harrington, Richard Harrington Jr., and Ron Skeddle

Rules of the Road for Entrepreneurs: A Map for a Successful Journey,
H. James Harrington and Shane P. Rogers

Organizational Change Management, H. James Harrington

Project Management, Review and Assessment, H. James Harrington and
William S. Ruggles

Cementing Customer Relationships, H. James Harrington and Thomas H. Carmody

Organizational Portfolio Management for Projects and Programs,
H. James Harrington, Richard Harrington Jr., and Ron Skeddle

Using Software Technology to Improve Organizational Performance,
H. James Harrington and William S. Ruggles

Effective Portfolio Management Systems

CHRISTOPHER F. VOEHL • H. JAMES HARRINGTON
WILLIAM S. RUGGLES

CRC Press
Taylor & Francis Group
Boca Raton London New York

CRC Press is an imprint of the
Taylor & Francis Group, an **informa** business

A PRODUCTIVITY PRESS BOOK

CRC Press
Taylor & Francis Group
6000 Broken Sound Parkway NW, Suite 300
Boca Raton, FL 33487-2742

© 2016 by Christopher F. Voehl, H. James Harrington, William S. Ruggles
CRC Press is an imprint of Taylor & Francis Group, an Informa business

No claim to original U.S. Government works

Printed on acid-free paper
Version Date: 20150817

International Standard Book Number-13: 978-1-4665-7253-9 (Paperback)

Visit the Taylor & Francis Web site at
http://www.taylorandfrancis.com

and the CRC Press Web site at
http://www.crcpress.com

I dedicate this book to a loyal friend whom I laughed with, sang with, traveled with, argued with, agreed with, learned from, and admired. He continuously encouraged me to do more, be better, and never stop learning. He taught me the importance of systems and convinced me that I should focus on process and performance improvement rather than just quality. Of all the quality gurus (Deming, Crosby, Juran, Ishikawa, and Masing), he was the one that had the biggest vision, the best understanding, and the most unique ideas. Unfortunately, he has passed away, as have all the other great gurus. I'm sure we all know who I am talking about. It is Armand (Val) Feigenbaum. Thanks for just being you.

H. James Harrington

I dedicate this book to my wife and business partner, Violeta Mercedes Granja de Ruggles ("Vivi Ruggles"). You have been an inspiration to me for the past 36 years and the reason for any successes I've had in my adult life. Tu éres mi vida!

William Ruggles

I dedicate this book to my amazing stepdaughter, Paige Carey. Thank you for the many ways you inspire me every day. It is such an honor to have a front row seat to witness your development into a bright and talented young woman. You have taught me so much about the important role of each parent in a child's life and I like to think I've taught you some things, too (aside from just video game and basketball skills). I know I still have much to learn as our little family grows and changes, but I'm excited to know that we will face these new experiences together. My life is forever changed for the better with you in it!

Christopher F. Voehl

Contents

Preface xi
Acknowledgments xiii
About the Authors... xv

Chapter 1 Overview of Organizational Portfolio Management 1

Introduction 1
Organizational Portfolio Development Cycle 6
    Phase I: Develop the Organizational Portfolio..................8
    Phase II: Create the OPM System Implementation Plan...11
    Phase III: Implement the OPM System..............................12
    Phase IV: Practical Applications of Project Change
    Management within the OPM System..............................13
Summary 14

Chapter 2 Phase I: Develop the Organizational Portfolio 15

Introduction 15
Developing the Organizational Portfolio: A Typical
Scenario...16
Thirteen Fundamental Terms...19
Activities for Phase I 21
    Activity #1: Assign an Individual and/or Team
    (Portfolio Development Leader) to Set Up and
    Manage the Development of the Organizational
    Portfolio...21
        Inputs...22
        Activities..27
        Outputs..29
    Activity #2: Classify the Business Cases Using a
    Qualitative, Quantitative, or Blended Model Based
    on the Potential Value 29
        Inputs... 30
        Activities..31
        Outputs..39

Activity #3: Prioritize the Projects and Programs
Based on the Resources Available...41
  Inputs... 43
  Activities.. 44
  Outputs...49
Activity #4: Select Projects and Programs for the
Portfolios and Assign a Portfolio Leader to Oversee
Them 49
  Inputs..50
  Activities...55
  Outputs... 56
Activity #5: Identify a Sponsor and Project Manager
for Each Project and Program... 56
  Inputs..58
  Activities...59
  Outputs...62
Activity #6: Define High-Level Project Milestones
and a Budget for Each Project and Program.....................62
  Inputs..63
  Activities...63
  Outputs... 68
Activity #7: Obtain Executive Approval for Each
Project and Program Including Its High-Level
Milestones and Budget... 68
  Inputs..70
  Activities...71
  Outputs...74
Summary 75
References 76

Chapter 3 Phase II: Create the OPM System Implementation Plan ... 77

Introduction 77
Activities for Phase II.. 80
  Activity #1: Prerequisites for OPM System
  Implementation Planning..81
  Activity #2: Establish the Project Management Office
  (PMO) 83
    Examples of Project Decentralization............................85

Activity #3: Assemble the OPM System
Implementation Team 86
Activity #4: Create the OPM System Implementation
Plan 88
   Sample OPM System Implementation Plan 89
References 97

Chapter 4   Phase III: Implement the OPM System 99

Introduction 99
OPM System Critical Success Factors 100
Activities for Phase III 100
Activity #1: Develop a Clear Vision of the
Organization's Strategic Goals and Objectives 101
Activity #2: Communicate the Change Agenda:
Goals, Objectives, Benefits, Risks, Rewards, and
Challenges 102
Activity #3: Identify Impacted Business Processes 104
Activity #4: Provide for Planning and
Implementation Phase Information Technology (IT)
Support 105
Activity #5: Develop Universal and Tailored Training ... 107
Activity #6: Develop Measurement and Reporting
Standards 109
Activity #7: Identify Risks and Technology Constraints ... 110
Activity #8: Schedule and Facilitate User Acceptance
Testing and End User Training 113
Activity #9: Develop Project/Portfolio Security and
Data Integrity Procedures 114
Activity #10: Implement the OPM System and
Report Progress 117
Summary 117
Reference 119

Chapter 5   Phase IV: Change Management—Practical
Applications within OPM 121

Introduction 121
Activities for Phase IV 121

Activity #1: Start at the Top...122
Activity #2: Create a Portfolio Enrollment and
Management Plan...124
Activity #3: Communicate the Rewards, Challenges,
Risk, and Consequences..126
Activity #4: Build Capacity within the Organization.....127
Activity #5: Integrate Risk Mitigation and Project
Planning...128
Activity #6: Plan for Sustained Results.............................130
Activity #7: Standardize the Portfolio Change
Management Approach...131
Summary...133

Appendix A: Project and OPM Definitions.................................. 135
Reference..145

Appendix B: Potential Project and OPM/PPM Resources.............. 147
CA PPM (formerly CA Clarity PPM) and CA PPM
as a Service...147
Features...147
Benefits...147
Daptiv...148
HP Project and Portfolio Management (PPM) Center...148
Microsoft™..149
Microsoft Office® Enterprise Project Management
(EPM) Solution..149
Microsoft Project Online: Cloud-Based Project
Portfolio Management Solution...150
Oracle Primavera: Professional Project
Management (P6) Software Suite...150
PD-Trak...150
Planview Enterprise..151
IBM Rational Focal Point..151

Appendix C: Sample Project Authorization Form (PAR)............... 155

Index.. 157

Preface

One of the biggest wastes occurring today in private and public organizations is the high percentage of failed projects and programs. Billions of dollars are flushed down the drain and millions of employee hours are wasted that should be directed toward improving our culture, environment, and advancements in science. Redeploying 25 percent of these wasted resources to medical research could result in huge gains in decreasing death rates and extending life expectancy.

There are many reasons why so many of our projects/programs have failed to deliver desired results. Some of them include

- Selecting the wrong projects
- Not staffing projects with the right personnel
- Lack of proper management attention
- Lack of a good sponsor
- Project resource overruns
- Greatly expanded scopes
- Poor project management
- Lack of coordination between projects
- Inadequate funding
- Projects are completed late, missing the window of opportunity
- Poor organizational change management
- Not providing adequate training
- Poor use of technology
- Not defining the best solution
- Setting wrong priorities
- Poorly defined goals and objectives
- Lack of understanding customer needs

This is just a partial list of things that can cause projects/programs to fail. I'm sure we could go on extending the list, but you should have the idea by now. Three of the biggest unresolved problems that organizations face are how to reduce the cycle time, the costs, and the failure rates of projects that have started, but do not produce an acceptable level of value added to the organization.

The processes and systems that we present in this book will not address all of the things that can cause projects/programs to fail; however, they will provide you with some guidance on how to select the *right* projects and programs. Also they will address how to apply the available resources to maximize results and potential success rates. The book will provide you as well with guidance on how to use Portfolio leaders to maximize the use of resources, minimize project overruns, and remove potentially failing projects early in the project/program cycle.

Acknowledgments

I would like to acknowledge all of the practical experience I gained while I was with Ernst & Young working on a number of portfolios of projects. They had developed a very systematic, documented, proven approach to managing a large portfolio of very challenging projects/programs to minimize their probability of failure and to maximize the profit Ernst & Young realized from working on the projects.

I also would like to acknowledge the unsung hero of this book, Candy Rogers, who coordinated and brought together the writings and thoughts of the three authors. She spent numerous hours combining the writings together into a book that has a continuous information flow, and edited the inputs to be sure we used the same terms and technology, thereby harmonizing the total manuscript.

H. James Harrington

About the Authors

H. James Harrington, PhD, is chief executive officer of Harrington Management Systems.

In the book, *Tech Trending* (Capstone, 2001), Dr. Harrington was referred to as "the quintessential tech trender." The *New York Times* referred to him as having a "… knack for synthesis and an open mind about packaging his knowledge and experience in new ways—characteristics that may matter more as prerequisites for new-economy success than technical wizardry… ." The author, Tom Peters, stated, "I fervently hope that Harrington's readers will not only benefit from the thoroughness of his effort, but will also 'smell' the fundamental nature of the challenge for change that he mounts." In 1991 President Bill Clinton appointed Dr. Harrington to serve as an Ambassador of Good Will. It has been said about him that "he writes the books that other consultants use."

Harrington Institute was featured on a half-hour TV program, *Heartbeat of America*, which focuses on outstanding small businesses that make America strong. The host, William Shatner, stated: "You [Dr. Harrington] manage an entrepreneurial company that moves America forward. You are obviously successful."

Currently, Dr. Harrington serves as the CEO for the Harrington Institute (Los Gatos, California). He also serves as the chairman of the board for a number of businesses, and is recognized as one of the world leaders in applying performance improvement methodologies to business processes. He has an excellent record of coming into an organization, working as its CEO or COO, resulting in a major improvement in its financial and quality performance.

Previously, in February 2002, Dr. Harrington retired as the COO of Systemcorp A.L.G., the leading supplier of knowledge management and project management software solutions when Systemcorp was purchased by IBM. Prior to this, he served as a principal and one of the leaders in the Process Innovation Group at Ernst & Young; he retired from Ernst & Young when it was purchased by Cap Gemini. Dr. Harrington joined Ernst & Young when it purchased Harrington, Hurd & Rieker, a consulting firm that Dr. Harrington started. Before that he was with IBM for over 40 years as a senior engineer and project manager.

He is past chairman and past president of the prestigious International Academy for Quality and of the American Society for Quality Control. He is also an active member of the Global Knowledge Economics Council.

Dr. Harrington was elected to the honorary level of the International Academy for Quality, which is the highest level of recognition in the quality profession.

He is a government-registered quality engineer, a certified quality and reliability engineer by the American Society for Quality Control, and a permanent certified professional manager by the Institute of Certified Professional Managers. He is a certified Master Six Sigma Black Belt and received the title of Six Sigma Grand Master. He has an MBA and PhD in engineering management and a BS in electrical engineering. Additionally, in 2013 Dr. Harrington received an honorary PhD from the Sudan Academy of Sciences.

His contributions to performance improvement around the world have brought him many honors. He was appointed the honorary advisor to the China Quality Control Association, and was elected to the Singapore Productivity Hall of Fame in 1990. He has been named lifetime honorary president of the Asia-Pacific Quality Control Organization and honorary director of the Association Chilean de Control de Calidad. In 2006, Dr. Harrington accepted the honorary chairman position of Quality Technology Park of Iran.

Dr. Harrington has been elected a fellow of the British Quality Control Organization and the American Society for Quality Control. In 2008, he was elected to be an honorary fellow of the Iran Quality Association and Azerbaijan Quality Association. He also was elected an honorary member of the quality societies in Taiwan, Argentina, Brazil, Colombia, and Singapore. He is listed in the Who's Who Worldwide and Men of Distinction Worldwide. He has presented hundreds of papers on performance improvement and organizational management structure at the local, state, national, and international levels.

Some of his recognitions include

The Harrington/Ishikawa Medal, presented yearly by the Asian Pacific Quality Organization, was named after H. James Harrington to recognize his many contributions to the region.

The Harrington/Neron Medal was named after H. James Harrington in 1997 for his many contributions to the quality movement in Canada.

Harrington Best TQM Thesis Award was established in 2004 and named after H. James Harrington by the European Universities Network and e-TQM College.

Harrington Chair in Performance Excellence was established in 2005 at the Sudan University.

Harrington Excellence Medal was established in 2007 to recognize an individual who uses the quality tools in a superior manner.

H. James Harrington Scholarship was established in 2011 by the ASQ (American Society for Quality) Inspection Division.

Dr. Harrington has received many awards, among them the Benjamin L. Lubelsky Award, the John Delbert Award, the Administrative Applications Division Silver Anniversary Award, and the Inspection Division Gold Medal Award. Other awards (by year) include

1996: The ASQC's Lancaster Award in recognition of his international activities.

2001: The Magnolia Award in recognition for the many contributions he has made in improving quality in China.

2002: Selected by the European Literati Club to receive a lifetime achievement award at the Literati Award for Excellence ceremony in London. The award was given to honor his excellent literature contributions to the advancement of quality and organizational performance.

2002: Awarded the International Academy of Quality President's Award in recognition for outstanding global leadership in quality and competitiveness, and contributions to IAQ as Nominations Committee chair, vice president, and chairman.

2003: The Edwards Medal from the American Society for Quality (ASQ). The Edwards Medal is presented to the individual who has demonstrated the most outstanding leadership in the application of modern quality control methods, especially through the organization and administration of such work.

2004: The Distinguished Service Award, which is ASQ's highest award for service granted by the Society.

2008: Awarded the Sheikh Khalifa Excellence Award (UAE) in recognition of his superior performance as an original Quality and Excellence Guru who helped shape modern quality thinking.

2009: Selected as the Professional of the Year.

2009: The Hamdan Bin Mohammed e-University Medal.

2010: The Asian Pacific Quality Association (APQO) awarded Harrington the APQO President's Award for his "exemplary leadership." The Australian Organization of Quality NSW's Board recognized Harrington as "the Global Leader in Performance Improvement Initiatives."

2011: The Shanghai Magnolia Special Contributions Award from the Shanghai Association for Quality in recognition of his 25 years of contributing to the advancement of quality in China. This was the first time that this award was given out.

2012: The ASQ Ishikawa Medal for his many contributions in promoting the understanding of process improvement and employee involvement on the human aspects of quality at the local, national, and international levels.

2012: The Jack Grayson Award. This award recognizes individuals who have demonstrated outstanding leadership in the application of quality philosophy, methods, and tools in education, healthcare, public service, and not-for-profit organizations.

2012: The A.C. Rosander Award. This is ASQ Service Quality Division's highest honor. It is given in recognition of outstanding long-term service and leadership resulting in substantial progress toward the fulfillment of the division's programs and goals.

2012: Honored by the Asia Pacific Quality Organization by being awarded the Armand V. Feigenbaum Lifetime Achievement Medal. This award is given annually to an individual whose relentless pursuit of performance improvement over a minimum of 25 years has distinguished himself or herself for the candidate's work in promoting the use of quality methodologies and principles within and outside of the organization of which he or she is a part.

Dr. Harrington is a very prolific author, publishing hundreds of technical reports and magazine articles. For the past eight years, he has published a monthly column in *Quality Digest Magazine* and is syndicated in five other publications. He has authored 40 books and 10 software packages.

William S. "Bill" Ruggles, MA, PMP, CQM, CSSMBB, CSM, is the managing partner of Ruggles & Ruggles, LLC based in New York City. He is an internationally recognized expert in managing and sponsoring projects, programs, and organizational portfolios with over 30 years of experience.

He has provided consultative, coaching, mentoring, training, and auditing support services to over 250 organizations and some 25,000 participants in 17 countries on 4 continents in 2 languages. These include many of the world's largest organizations in the fields of healthcare, biopharmaceutical, biotechnology, medical devices, information technology, global communication technologies, insurance, banking/financial services, high tech manufacturing, engineering, shared services, and the public sector for both the state and federal governments.

Ruggles has been an adjunct professor of project management at Stevens Institute of Technology (Hoboken, New Jersey) since 2000 and has also lectured at William Paterson University (Wayne, New Jersey), the New Jersey Institute of Technology (Newark), ITESM/Monterrey Tech (Monterrey, Mexico), Universidad del Valle (Ciudad de México, Mexico), Universidad San Francisco de Quito (Quito, Ecuador), PT Telekom RISTI Center (Jakarta, Indonesia), Fundación Chile (Región Metropolitana, Chile), Makro Education Center (Istanbul, Turkey), Boston University (Boston, Massachusetts), and UTStarcom University (Beijing, China).

His practice specialties include delivering quality and continuous process improvement outcomes, and enhancing project and program performances via the development of organizational, interpersonal, and individual proficiencies and competencies among those who sponsor, manage, and work on project, program, and portfolio teams.

Ruggles has been certified as a Project Management Professional (PMP) by the Project Management Institute, as a Certified Quality Manager (CQM) by the American Society for Quality, as a Certified Six Sigma Master Black Belt (CSSMBB) by the Harrington Institute, and as a Certified Scrum Master (CSM) by the Scrum Alliance. He holds an MA from Columbia University and a BS from the University of Connecticut and is fluent in Spanish.

Ruggles served as the president of PMI's (Project Management Institute) New Jersey Chapter (1990–1992) and served on PMI's international board of directors as its vice president of technical activities and president of the PMI Educational Foundation (1994–1995), as an ex-officio member of the board of directors (1996), as president and chief operating officer (1997), and as board chair (1998). He also was a recipient of both the Distinguished Contribution Award and the Linn Stuckenbruck Person-of-the-Year Award in 1996. More recently, he served as the vice president of administration for PMI's New York City chapter (2011–2012) and as an advisor for its New Jersey chapter (2013–2014).

Christopher F. Voehl is the president of Seven Sigma Tools (Tallahassee, Florida), a company focused on accelerating client performance, customer satisfaction, systems deployment, project management, business development, process optimization, and continuous improvement services.

Voehl has over 20 years of technical, consulting, and executive management experience spanning multiple disciplines in a variety of industries, specializing in process optimization and client value delivery for business services, human capital consulting firms, and nonprofits, while helping service industry and healthcare organizations adapt continuous improvement methodologies including TQM, ISO 9001, and Lean Six Sigma.

Recently, Voehl has focused on government sector quality improvement initiatives in the United States and the Caribbean, while also leading global call center transformation initiatives internationally in Australia and Saudi Arabia, providing clients with detailed assessments of strategy/ vision alignment, call center quality, organizational alignment, customer satisfaction, process optimization, and technology utilization. Recent contact center optimization projects focused on implementation of new technology, profit optimization, and recruitment process outsourcing.

Voehl managed the development of the Lean Six Sigma executive education curriculum for Nova Southeastern University (NSU), Fort Lauderdale, Florida. In addition, he has served as a course instructor and mentor at NSU and the University of Central Florida (Orlando). These programs have certified hundreds of White Belts, Yellow Belts, Green Belts, and Black Belts since 2009.

Voehl has been an American Society for Quality (ASQ) member since 1995, is a certified Lean Six Sigma Master Black Belt, and is the author of several articles and publications on quality, technology, and business analysis including the book *Making the Case for Change: Using Effective Business Cases to Minimize Project and Innovation Failures* (CRC Press, 2014) and white papers for the Project Management Institute (PMI) including *Organizational Change Management and the Model for Sustainable Change* (January 2015) with coauthors H. James Harrington and Frank Voehl.

1

Overview of Organizational Portfolio Management

INTRODUCTION

I have too few resources and too many new projects. How can I possibly keep the present commitments that are being made by our sales force and still assign the resources required to support the new projects that are going to drive our future? I need to keep our delivery commitments to our present customers as my top priority or we won't have customers when the new projects are completed.

We hear this comment over and over and, when we don't hear it, it is because they are thinking it, but don't want to say it. New projects are the lifeblood of most organizations. Without them the organization has little or no future.

Once we opened up the world as our customers, we also opened up organizations around the world as our competition. This has resulted in the highly competitive research activities that have resulted in extremely short product cycles. Many of our best-known and most productive organizations depend upon technology as their edge against competition. However, in today's environment, technology can be picked up and transferred around the world in a matter of hours resulting in very little competitive advantage. Where technology and the latest software applications used to be considered a competitive advantage, it now is a requirement to keep from becoming noncompetitive. As a result, there has been a great deal of pressure to reduce project cycle times and greatly improve the percentage of projects that are completed successfully. The organizations that are the most successful today are the organizations that can create new concepts on demand, minimize the time from concept to delivery, and have the highest percentage of projects that are successful.

Many people are promoting the idea that we learn from our failures. That's a good concept, but if it's the one you are using, let's hope you don't spend a significant portion of your resources learning instead of succeeding.

Once we increased the emphasis on innovation, which has resulted in more projects being introduced into our systems with increased requirements for implementing them in shorter periods of time, the importance of effective management of these projects became a key element for successful organizations. With this increased emphasis on efficient and effective project management activities, the concept of Organizational Portfolio Management became a key element in an organization's competencies. The purpose of this book is to highlight effective, proven approaches to maximize the quantity and caliber of the innovative concepts that successfully complete the process from business plan to external positive results based on the resource limitations within the organization. This requires a continuous focus on optimizing the resource usage consumed by the Organizational Portfolio that is going on within the organization.

In order to eliminate any confusion, we are employing the following definitions of key terms used. (Appendix A is a complete list of definitions and acronyms that are commonly used throughout this book.)

- **Business Case:** A business case captures the reason for initializing a project or program. It is most often presented in a well-structured written document, but, in some cases, also may be in the form of a short verbal argument or presentation. The logic of the business case is: Whenever resources, such as money or effort, are consumed, they should be in support of a specific business need or opportunity.
- **Manager:** A manager is an individual who accomplishes an assigned task through the use of other individuals to whom the work is delegated.
- **Organizational:** This refers to those activities, projects, programs, processes, and systems that apply to the total organization, not just one or two departments or units.
- **Organizational Portfolio:** A collection of the organization's projects, programs, subportfolios, and operations managed as a group to achieve strategic objectives. Projects and programs may work independently, but they are often linked to the organization's strategic plan. An individual organizational portfolio can contain all of the organization's active projects and/or programs or a portion thereof.

- **Organizational Portfolio Management (OPM):** The combined coordination and management of all the active projects/programs to maximize the value they add to the organization by continuously monitoring their progress, prioritizing work, and allocating resources. It refers to the combined activity of all the active portfolios and independent projects going on within the organization and the improvement of the organization's capability linking Project/Program/Portfolio Management with organizational facilitators (structural, cultural, technological, Human Resources practices) to support strategic goals. To apply this methodology, organizations must first measure existing capabilities, then plan and implement the improvements.
- **Organizational Portfolio Manager:** The individual that is assigned to manage the organization's Portfolio leaders. The individual is held responsible for Organizational Portfolio Management activities. In organizations that have a project office, this individual is often referred to as the project office manager.
- **Organizations:** Systematic arrangements of entities (people, departments, companies, divisions, teams, agencies, etc.) aimed at accomplishing a purpose, which may involve undertaking projects. They often are documented in an organization chart that shows the relationships of the individual organization to the total organization.
- **Portfolio Leader:** A senior project manager qualified and appointed to lead multiple concurrent programs and/or projects. Portfolio leader roles are typically project managers with years of demonstrated success organizing and managing projects/programs. Some organizations call this individual a Portfolio manager. They serve as one part of the total Organizational Portfolio Management activities.
- **Portfolio Steering Committee:** An executive committee that is responsible for overseeing all the active projects/programs. All major changes in timing, funding, and an output from the project process need to be approved by this committee.
- **Process:** It is a series of interrelated actions and/or tasks performed to create a prespecified product, service, or result. Each process is comprised of inputs, outputs, tools, and techniques, with constraints (environmental factors), guidance, and criteria (organizational process assets) taken into consideration.
- **Program:** A group of related projects, subprograms, and program activities managed in a coordinated way to obtain benefits

not available from managing them individually. May include work outside the scope of projects within it. A program will always have projects.

- **Program/Project Management:** The application of tasks/tools/ techniques and skills/knowledge to meet program requirements, to obtain benefits and control not available from managing them individually. It is harmonizing projects and program components, controlling interdependencies to achieve benefits outlined in the business case and value proposition.
- **Project:** A temporary endeavor undertaken to create a unique product or service.
- **Project and Portfolio Management Systems:** This is software used to enable centralized management of processes, methods, and technologies by Portfolio leaders, project managers, and Project Teams to concurrently analyze and manage all proposed and active projects.
- **Project Life Cycle:** This is a collection of generally sequential project phases whose names and number are determined by the control needed in the organization or organizations involved in the project.
- **Project Management Office:** This is an official department that is usually recognized on the organization's organization chart. This is the department managed by an individual that has a number of, if not all, the project managers and Portfolio leaders reporting to him or her. It is the department that is responsible for defining the system that is used to manage portfolios, projects, and programs.
- **Project Manager:** An organizational employee, representative, or consultant appointed to prepare a project/program and who plans and organizes the resources required to complete a project/program, prior to, during, and upon closure of the project/program life cycle. Note: *Project manager* is also the term used for individuals who are managing programs.
- **Project Portfolio:** A centralized collection of independent projects or programs that are grouped together to facilitate their prioritization, effective management, and resource optimization in order to meet strategic organizational objectives.
- **Project Team:** This is a team of people who are responsible for designing, developing, implementing, and measuring the effectiveness of an approved project/program. The majority of the team usually will report to the Project Team manager.

- **Project Team Manager:** This is an individual who is truly accountable for the success or failure of a specific project/program. He/she usually will have many, if not all, of the people working on the project/program assigned directly to him/her. The Project Team manager will be responsible for getting people from other organizations to work on projects as they are needed. The project manager's job is to monitor how successfully the Project Team manager is performing his/her assigned responsibilities. Projects that do not have project managers assigned to them will be completely managed by the Project Team manager.
- **Subportfolio Leader:** In some cases, due to the complexity and number of individual projects going on in a specific area, a major portfolio may be divided into subportfolios with a Portfolio leader assigned to each of them. For example, an organization's R&D activities may be directed at a number of different product types and, in each of the product types, a number of different projects are being actively pursued. In these cases, each specific type of product may have a Portfolio leader coordinating/managing all the individual projects/programs. Some organizations classify these individuals as Subportfolio leaders related to that product type. To reduce confusion and to simplify the manuscript, we are using the term *Portfolio leader* to represent both the Portfolio leader and the Subportfolio leader.
- **Value Proposition:** A short statement that describes the tangible results/value a decision maker can expect from implementing the recommended course of action and its benefit to the organization.

Now, let's set the stage for well-managed, progressive organizations functioning in today's global work environment. They have realized that innovation is the key to a successful future. They have initiated communication and training systems to prepare their staff to be highly creative individuals who are prepared to take risks to make the organizations more successful. The organization is flooded with creative and innovative ideas from marketing, sales, product engineering, finance, information technology, product engineering, manufacturing engineering, and research and development. Each of these functional groups has a value proposition prepared for it that screened out the ideas that did not provide acceptable value-added content to the organization. These screened, innovative approaches have had business plans prepared for them by an independent group that indicates that they have a high probability of adding value to the organization. Many of them are short-term, easy-to-implement concepts that are refinements to the current activities. Others will require a long-term investment with the known risks associated with them. The ones that were successful in completing the business plan review are all aligned with the organization's strategic plan, future visions, and the organization's values. The problem now rests with selecting the projects/programs that will provide the maximum value both short range and long range to the organization based upon the resources that can be made available to support the projects/programs. While all of the projects/programs that successfully completed the business plan review have the potential of providing an acceptable added-value level to the organization, they cannot all be approved due to a limit on the money, staff, and facilities that are available. As a result, the organization needs to select an Organizational Portfolio that maximizes the values that can be created from the implementation of the projects and programs that make up the Organizational Portfolio for the organization.

ORGANIZATIONAL PORTFOLIO DEVELOPMENT CYCLE

This book is designed to take the reader through the complete project/program management cycle from the submittal of the proposed projects/programs to the management of their implementation. To accomplish this, an effective, proven, four-phase Organizational Portfolio Management (OPM) System is defined. This is not the only OPM System that can be used, but it is one that we know is both efficient and effective.

- Phase I: Developing the Organizational Portfolio is selecting the right mix of projects/programs that will make up the Organizational Portfolio based upon the resource limitations that are present and the risks involved that are related to the individual projects and programs. Figure 1.1 is a box flowchart that represents the process flow for Phase I.

  The cycle starts with information related to all of the proposed projects/programs and all the currently active projects/programs being provided to a Portfolio Development Team. The team analyzes and evaluates each of these to determine the value added they represent to

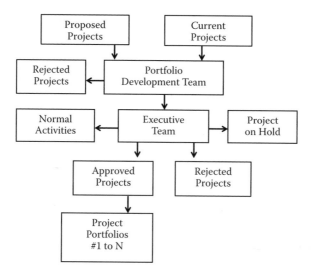

FIGURE 1.1
Organizational Portfolio development cycle.

the organization, the related risks, and the resources required compared to resources available, to determine which project/program should be rejected and which should be recommended to the Executive Team to have resources applied to them. The Executive Team analyzes the recommendations made by the Portfolio Development Team and makes decisions on whether the projects/programs will be approved, rejected, or put on hold. Often some of the proposed projects/programs do not require additional resources because they are already part of the function's normal activities and budget. The approved projects/programs are grouped into a manageable Organizational Portfolio.

- Phase II: Create the OPM System Implementation Plan. This phase develops a plan for establishing and managing the OPM System to minimize the resources consumed and cycle time to complete the projects, and increase the ability of the projects to meet their projected value-added content to the organization.
- Phase III: Implement the OPM System. This phase focuses on the complexity of managing an Organizational Portfolio and keeping it aligned with the organization's vision goals and objectives. It requires the development of an effective rewards and recognition system that supports the changes brought about by the project/program. It focuses attention on the needs for training the individuals who are impacted by the project/program prior to its implementation.

- Phase IV: Practical applications of Project Change Management within the OPM System. This phase focuses on the difficulties and how to overcome them related to the continuous changing environment and project requirements that are encountered as projects are developed and implemented in today's demanding conditions.

Before we get into some detailed discussion related to each of these four phases, there are some additional definitions that need to be provided. They are primarily related to the OPM system structure.

- **Advocate:** The individual or group that wants to achieve a change, but lacks the power to sanction it.
- **Change Agent:** The individual or group responsible for facilitating the implementation of the change.
- **Impacted Individuals:** These are the individuals whose activities will be directly affected by the output from the project/program when the project/program is completed.
- **Initiating Sponsor:** The individual or group with the power to initiate or to legitimize proposed project or program related to all of the affected people in the organization.
- **Sponsor:** The individual or group with the power to sanction or legitimize projects.
- **Sustaining Sponsor:** The individual or group that can use their logistics, their economic and/or political proximity to the individuals affected by the change, to convince them that they should support, help implement, and comply with the project/program.

Phase I: Develop the Organizational Portfolio

The activities that take place during this phase are some of the most difficult activities that the organization performs. This is the phase where it is very difficult to set up win–win scenarios for everyone involved in it. Some projects that individuals may have been working on for months if not years that have been projected to add value to the organization may need to be dropped in favor of other projects. Some of the typical reasons why projects are dropped include:

- Miscalculations in the original justifications.
- Changes in the marketplace.

- Too many changes already taking place in the portion of the organization that would be affected.
- Projected technology advances will render the project obsolete before significant return on investment will be realized.
- The lack of available resources with the proper skills.
- The lack of sufficient financial resources.
- The lack of a legitimate sponsor.
- The risks associated with the project are too great compared to other projects that are available.
- Another project will accomplish the same results with fewer resources invested.
- Identification of potential patent infringements.
- The lack of a champion for the project.
- The project is not in line with the short-term strategic goals.
- The project is not in line with the organization's culture.
- A related dependency will not be in place in time to support the project.
- Political disagreements between members of the executive team.
- Other activities that are going on that would eliminate the need for the project.

Phase I is made up of seven different activities that include:

Activity #1: Assign an individual and/or team (Portfolio Development leader) to set up and manage the development of the Organizational Portfolio. The major outputs from Activity #1 include:

- A set of criteria aligned to the organization's mission designed to provide guidance related to certifying and ranking the business cases.
- The Portfolio Development leader has been selected and is on board. This includes having prepared a job description for him/her.
- Portfolio Development Team member(s) selected and onboard. This includes preparing job descriptions for the team members of the Portfolio Development Team.

Activity #2: Classify the business cases using a quantitative, qualitative, or blended model based on potential value. The major outputs from Activity #2 include:

- An Executive Committee review and approved rank-ordered list of projects and programs based upon their potential value to the organization.

Activity #3: Prioritize the projects and programs based on the resources available. The major outputs from Activity #3 include:

- A list of new, resource-leveled, rank-ordered projects/programs that will be included in the organization's active projects.
- A list of new, rank-ordered projects/programs with no resources assigned and are deferred.
- A list of projects/programs that are *not* recommended for implementation and, therefore, are dropped.
- A written description for all approved projects and projects put on hold that includes the goals, objectives, and resource requirements.
- Documented performance and projected resource requirements for each project/program.

Activity #4: Select projects and programs for the portfolio(s) and assign a Portfolio leader to oversee it/them. The major outputs from Activity #4 include:

- A list of resource-leveled, rank-ordered proposed projects/programs in the portfolio with each one's assigned Portfolio leader.
- A list of initiating sponsors for all current and approved projects/ programs.
- A list of Portfolio leaders with subportfolios and projects/programs assigned.
- A list of all current active projects/programs/portfolios.

Activity #5: Identify a sponsor and a project manager for each project and program. The major outputs from Activity #5 include:

- A list of projects and programs that require project management with the name of the project manager responsible for the project/program.
- A list of the project managers with the names of the projects that are assigned to each.
- A list of projects that do not require a project manager to be assigned to them.

Activity #6: Define high-level milestones and a budget for each project and program. The major outputs from Activity #6 include:

- High-level budget set for each project.
- Key management checkpoints are defined and documented.

- The Project Team is defined and assigned.
- Executive summaries are prepared for each proposed project/program.

Activity #7: Obtain executive approval for each project and program including its high-level milestones and budget. The major outputs from Activity #7 include:

- List of approved projects/programs with their final executive summaries, start and end dates, and budget for each one.
- An approved list of portfolios with project manager, project team manager, and sustaining sponsors for each project/program.
- A formal project report that summarizes their activities and assumptions.
- Minutes of the executive committee review and decisions.
- Updated set of operating procedures related to the process of selecting the projects/programs that will be included in the Organizational Portfolio.
- Project team manager assigned to each approved project/program.
- Start and end dates and a budget for each project.
- Key checkpoints for each project or program, if applicable.
- Executive summary for each proposed project/program.

Each of the above seven activities are organized initially by defining the inputs that are required to start the activity. It then discusses in detail the tasks that are required to perform the activity. The activity is finalized by defining and describing the outputs from the activity. At the end of this phase, the project/programs that successfully complete Activity #7 are ready to be turned over to the project manager and the Portfolio leader.

Phase II: Create the OPM System Implementation Plan

At the completion of Phase I, all the potential and active programs/projects have been analyzed and decisions have been made on which of them would be activated or remain active. Preliminary budgets for the newly activated programs/projects have been approved. Relevant projects/programs have been grouped together into portfolios with Portfolio leaders assigned. Now the task that faces the organization is to develop an OPM System that will provide an effective guidance/management for the approved projects/programs. To accomplish this, Phase II has been divided into the following four activities.

Note: For organizations that have already established an OPM System that is operating effectively, many of these activities can be minimized or dropped. For those organizations that already have an OPM System in place, we still recommend that Phase II be used to review the present system to define how it can be made more efficient and effective.

> **Activity #1:** Prerequisites for OPM System Implementation Planning.
> **Activity #2:** Establish the Project Management Office (PMO).
> **Activity #3:** Assemble the OPM System Implementation Team.
> **Activity #4:** Create the OPM System Implementation Plan.

Phase III: Implement the OPM System

As the name for Phase III implies, this phase is directed at preparing the organization to implement the OPM System. All too often creative innovative concepts are developed and effectively documented that fail due to poor implementation. The concept developers simply throw the concept over the wall and hope someone else catches the ball and runs with it. This is a sure way to minimize the effectiveness of an otherwise excellent system/concept. To minimize the possibility of this happening, Phase III is divided into the following 10 activities:

> **Activity #1:** Develop a clear vision of the organization's strategic goals and objectives.
> **Activity #2:** Communicate the change agenda: goals, objectives, benefits, risks, rewards, and challenges.
> **Activity #3:** Identify impacted business processes.
> **Activity #4:** Provide for planning and implementation phase information technology (IT) support.
> **Activity #5:** Develop universal and tailored training.
> **Activity #6:** Develop measurement and reporting standards.
> **Activity #7:** Identify risks and technology constraints.
> **Activity #8:** Schedule and facilitate user acceptance testing and end user training.
> **Activity #9:** Develop project/portfolio security and data integrity procedures.
> **Activity #10:** Implement the OPM System and report progress.

Phase IV: Practical Applications of Project Change Management within the OPM System

There have been hundreds of books written and millions of hours devoted to defining an effective way to manage projects. The worldwide leader in this scientific endeavor is the nonprofit association Project Management Institute (PMI). It has done a superb job of defining the Project Management body of knowledge in its publication, *A Guide to the Project Management Body of Knowledge (PMBOK Guide)* (PMI, 2009), which addresses nine key knowledge areas:

1. Project Integration Management
2. Project Scope Management
3. Project Time Management
4. Project Cost Management
5. Project Human Resource Management
6. Project Quality Management
7. Project Communication Management
8. Project Risk Management
9. Project Procurement Management

Each of these key knowledge areas are further divided into processes and activities that explain how, when, and where to address them. In spite of PMI's thorough research, one of the areas that we feel has been not fully developed is how project management needs to focus on Organizational Change Management. Due to Dr. Harrington's strong feeling about this missing area, he wrote a book entitled *Project Change Management: Applying Change Management to Improve Projects* (McGraw-Hill, 1999). This book documented the approaches used by Ernst & Young that resulted in its successfully completing over 90 percent of all their projects. This book suggested that a 10th project management knowledge area—Project Change Management—be added to the *PMBOK*. In 2012, PMI added a 10th knowledge area, but it was not Organizational Change Management, it was Project Stakeholder Management.

Although PMI has recognized the importance of Project Change Management and it has started to integrate it into parts of the other 10 knowledge areas, to date the integration into actual practice has been slow in taking place. As a result, we are recommending a final phase—Phase IV: Practical applications of Project Change Management within OPM. To accomplish this, we are suggesting that you apply the following seven activities:

Activity #1: Start at the top.
Activity #2: Create a Portfolio Enrollment and Management Plan.
Activity #3: Communicate the rewards, challenges, risk, and consequences.
Activity #4: Build capacity within the organization.
Activity #5: Integrate risk mitigation and project planning.
Activity #6: Plan for sustained results.
Activity #7: Standardize the Portfolio change management approach.

SUMMARY

This chapter has focused on how to analyze the many proposed projects/programs and compare them to the resources that are available in order to select the ones that provide the maximum value to the organization. It also took into consideration the projects that were presently active to determine which if any of them should be dropped in favor of a proposed project or program. This chapter also took the reader through the activities necessary to analyze the approved projects and group them together in a homogeneous portfolio in order to gain additional assurance that maximum value will be obtained from the resources utilized on the projects/programs. The selection process is based on one simple phrase. It is: "The best time to stop a program that's going to fail is before it starts."

2

Phase I: Develop the Organizational Portfolio

The best plans of mice and men are only as good as the way they're implemented.

H. James Harrington

The sated appetite spurns honey, but to a ravenous appetite even the bitter is sweet.

Proverbs 27:7

INTRODUCTION

The object of the activities that make up this chapter is to develop the Organizational Portfolio by evaluating the set of proposed business cases (each proposed business case was already aligned with a value proposition) to determine which of them will be selected as a project or program to be approved (or not) and included in the portfolio, either to be launched for the first time or, if one has already been established, to continue in the next time period.

Before proceeding any farther, let's understand three of the fundamental terms we will be using throughout the remainder of the book.

1. *Portfolio*: A centralized collection of independent projects or programs that are grouped together to facilitate their prioritization, effective management, and resource optimization in order to meet strategic organizational objectives.

2. *Value Proposition*: A document based on a review and an analysis of the benefits, costs, and value that an organization or an individual project or program can deliver to its internal/external customers, prospective customers, and other constituent groups within and outside the organization. It is also a positioning of value, where Value = Benefits – Cost (where cost includes risk).

3. *Business Case*: This is an evaluation of the potential impact a problem or opportunity has on the organization to determine if it is worthwhile investing the resources to correct the problem or take advantage of the opportunity. An example of the results of the business case analysis of the software upgrade could be that it would improve the software's performance as stated in the value proposition, but (a) it would decrease overall customer satisfaction by an estimated three percentage points, (b) require 5 percent more task processing time, and (c) reduce system maintenance costs only $800 a year. As a result, the business case did not recommend including the project in the portfolio of active programs. Often the business case is prepared by an independent group, thereby, giving a fresh, unbiased analysis of the benefits and costs related to completing the project or program.

DEVELOPING THE ORGANIZATIONAL PORTFOLIO: A TYPICAL SCENARIO

The term *business case* is frequently used as part of the annual budget cycle for the organization. To help you understand the complexity of defining which business cases are included in the Organizational Portfolio, we present the following organizational structure for a "typical" manufacturing organization (Figure 2.1).

The organization is in the middle of its annual budgeting cycle. Estimates on the resources required to support the previously approved products, programs, and projects have already been submitted and they are running about 18 percent over projected budgets already. In addition, each function has submitted new business cases (see the series: Little Big Book—*Making the Case for Change: Using Effective Business Cases to Minimize Project and Innovation Failures* (CRC Press, 2014)) that they would like to initiate

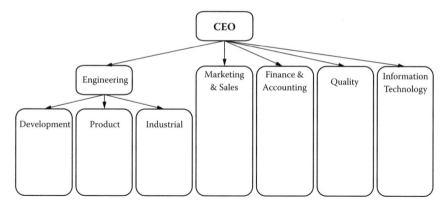

FIGURE 2.1
Typical manufacturing organizational chart.

since they would (or should) fulfill one of the identified value propositions (see the series: Little Big Book—*Maximizing Value Propositions to Increase Project Success Rates* (CRC Press, 2014)). As a result, a stack of approximately 125 unranked business cases are sitting on the CEO's desk awaiting approval. A quick review of the business cases indicates that it would require 35 percent more available resources to implement all of these proposed improvements.

Now, among them is a set of these proposed business cases from **Development Engineering**, all of which would make excellent technical research papers, but some of them also are directly aligned with the organization's mission and vision statement.

There is another set from **Product Engineering** for new products that have potential to generate additional revenue for the company. However, the CEO fears that only some of these proposals will actually be profitable and the others will be "money losers" for the organization.

Industrial Engineering wants to automate the DASD (direct access storage device) line, but **Production Control** has a proposal in to subcontract and offshore that same secondary storage device line to China.

Marketing and Sales wants to implement a new, enterprise-wide Customer Relationship Management (CRM) system to find, attract, and win new clients; nurture and retain those the company already has; entice former clients to return; and reduce the costs of marketing and client services. However, this Marketing and Sales initiative would require heavy investment in Information Technology resources that may be better utilized on other undertakings.

Finance and Accounting wants to convert to an automated activity-based costing system to reduce the cost of order processing activities.

Quality has proposed redesigning the order processing process.

Finally, it seems like just about everyone has some work that they want the **Information Technology** Department to do for them to make their respective jobs easier. Besides, the CIO (chief information officer) wants to upgrade the enterprise to the cloud version of latest office productivity software suite.

Some of these business cases indicate that there are *no* additional resources required to implement them, but the CEO questions if these resources would be better utilized doing something else that would create more value added to the organization. The CEO realizes that if *all* these initiatives were approved, the organization wouldn't have enough resources to manufacture the present product lines that are paying everyone's salaries.

Because the organization needs to reduce costs and create maximum value with these resources, some of the projects that are presently underway, but are underperforming, may be of less value than the ones that are currently being proposed. "If so," he wonders, "which current project(s) should be dropped to free up resources to implement the more profitable new projects or programs?" Based on past experience, the CEO knows that some functions tend to underestimate the resources required for a project and overestimate the value that the project will create. Other functions do just the opposite by overestimating the resources required and underestimating the project's value to the organization.

Because almost all projects take longer to implement and cost a great deal more than estimated, a given business case based on a particular value proposition may sound very straightforward and optimistic, but then, along come the "add-ons" and systems interfaces. It's not unusual to find that, once a project has been approved for, say, $1 million, halfway through the schedule, the CEO is informed that it's going to fail unless an additional $100,000 is invested in it. So, he is pretty much obligated to spend the additional resources to salvage the project (the classic "sunk cost" dilemma). In fact, this seems to be more the rule than the exception in recent years. As a result, budget overruns and schedule delays are prevalent on most projects.

Returning to our "typical" Organizational Portfolio scenario, there are nine questions that the CEO should ask at this point in the portfolio development process:

1. To whom should I give this stack of business case proposals who will give an unbiased evaluation of which ones should be included in the organization's active portfolio of projects and programs?
2. How should these initiatives be prioritized?
3. How will the stakeholders react to the implementation of the approved initiatives?
4. How many resources are available to support the newly proposed initiatives?
5. Do we have the right skills to implement the approved initiatives and, if not, how do we find them?
6. Who is going to be managing each of the approved initiatives?
7. What is the schedule for implementing each of them and how do they relate to the work activities that are already going on within the organization?
8. What are the interrelationships and interdependencies between the approved initiatives?
9. Which initiatives need to be grouped into a portfolio or group of portfolios and which ones can be handled as a part of the functional department's normal activities?

In this chapter, we provide answers to these nine questions the CEO should be asking when focusing on the development initiatives (projects and programs).

THIRTEEN FUNDAMENTAL TERMS

Before proceeding, let's clarify 13 more fundamental terms that are used in the remainder of this book:

1. *Mission Statement*: The stated reason for the existence of the organization. It is usually prepared by the CEO and key members of the Executive Team and succinctly states what they will achieve or accomplish. It is typically changed only when the organization decides to pursue a completely new market.
2. *Policy*: A principle or rule to guide decisions and achieve rational outcomes; an intent to govern that is implemented as a procedure. Policies are generally adopted by the Board of Directors or senior governance body within an organization, whereas, procedures are

developed and adopted by senior and middle managers. Policies can assist in both subjective and objective decision making.

3. *Vision Statement*: It provides a view of the future desired state or condition of an organization. (A vision should stretch the organization to become the best that it can be.) The vision statement provides an effective tool to help develop objectives.

4. *Value*: The basic beliefs or principles upon which the organization is founded and which make up its organizational culture. They are prepared by top management and are rarely changed because they must be statements that the stakeholders hold and depend on as being sacred to the organization.

5. *Strategy*: It defines the way the mission will be accomplished. Using a well-defined strategy provides management with a thought pattern that helps them better utilize equipment and direct resources toward achieving specific goals; e.g., "The company will identify new customer markets within the United States and concentrate on expanding markets in the Pacific Rim countries."

6. *Critical Success Factors*: These are the key things that the organization must do extremely well to overcome today's problems and the roadblocks to meeting the vision statements.

7. *Business Objectives*: Business objectives are used to define what the organization wishes to accomplish, often over the next 5 to 10 years.

8. *Organizational Goals*: They document the desired, quantified, and measurable results that the organization wants to accomplish in a set period of time to support its business objectives; e.g., increase sales at a minimum rate of 12 percent per year for the next 10 years with an overall average annual growth rate of 13 percent. Goals should be specific rather than general so that there is no ambiguity.

9. *Strategic Business Plan*: This plan focuses on what the organization is going to do to grow its market share. It is designed to answer the questions: What do we do? And how can we beat the competition? It is directed at the product.

10. *Organizational Master Plan*: The combination and alignment of an organization's Business Plan, Strategic Business Plan, Combined Performance Acceleration Management (PAM) Plan, and Annual Operating Plan.

11. *Business Plan*: A formal statement of a set of business goals, the reason they are believed to be obtainable, and the plan for reaching these goals. It also contains background information about

the organization and/or services that the organization provides as viewed by the outside world.

12. *Initiating Sponsor:* Individual/group that has the power to initiate and legitimize the change for all of the affected individuals.

13. *Sustaining Sponsor:* The individual/group that has the political, logistical, and economic proximity to the individuals affected by a new project/activity.

ACTIVITIES FOR PHASE I

Phase I (*Develop the Organizational Portfolio*) is made up of the following seven activities:

Activity #1: Assign an individual and/or team (Portfolio Development leader) to set up and manage the development of the Organizational Portfolio.

Activity #2: Classify the business cases using a quantitative, qualitative, or blended model based on each one's potential value.

Activity #3: Prioritize the projects and programs based on the resources available.

Activity #4: Select projects and programs for the portfolio(s) and assign a Portfolio leader to oversee it/them.

Activity #5: Identify a sponsor and a project manager for each project and program.

Activity #6: Define high-level milestones and a budget for each project and program.

Activity #7: Obtain executive approval for each project and program including its high-level milestones and budget.

Below, we define each activity, in sequence, and identify each one's respective inputs, activities, and outputs.

Activity #1: Assign an Individual and/or Team (Portfolio Development Leader) to Set Up and Manage the Development of the Organizational Portfolio

For a **small organization** (10 or fewer combined projects and programs,), this "entity" could be an *individual* with the right leadership credentials and characteristics. For a **medium-sized organization** (10–25 combined projects and programs,), this could be *either one or two individuals or a department led by an individual* with the right leadership credentials and characteristics. For a **large organization** (more than 25 combined projects and programs), this should be a *department led by an individual* with the right leadership credentials to be able to handle the demands adequately. Usually this assignment is given to one of the established functions to coordinate like Finance, Project Management, or Project Engineering.

Here are the inputs, activities, and outputs for Activity #1 to recruit and assign the appropriate entity or entities to develop the Organizational Portfolio(s):

- Input(s):
  - The organization's mission statement and strategic plan
  - OPM mission statement from the Executive Team
  - Portfolio Development leader credentials and characteristics
  - Portfolio Development Team role and responsibilities (if appropriate)
  - Portfolio Development Team member credentials and characteristics (if appropriate)
- Activities:
  - Talent recruitment and acquisition
  - Candidate interviewing for the Portfolio Development leader
  - Human Resource selection for the Portfolio Development leader assignment
  - Interviewing and selection of individuals for the Portfolio Development Team assignment
- Output(s):
  - A set of criteria aligned to the organization's mission to classify and rank the business cases
  - A Portfolio Development leader is selected and on board
  - Portfolio Development Team member(s) are selected and on board (if appropriate)

Inputs

- The organization's mission statement and its strategic plan
  These two are key inputs into evaluating and prioritizing projects/programs. All of the approved projects and programs should directly support the organization's mission statement that defines the kind

of business the organization was formed to be involved in. They also should be directly in line with the organization's strategic plan. The strategic plan provides guidance to the total organization related to how the Executive Team and the Board of Directors desire the organization to evolve over the near-term future. Usually this is a 5- to 10-year period. Projects and programs that are not aligned with the organization's mission statement should not be considered for implementation unless there is a change to the mission statement. Projects and programs that are in line with or support the organization's mission statement, but are not directly in line with the organization's strategic plan, can be considered candidates to become part of the organization's portfolio of active projects/programs. However, these projects and programs have a far lower chance of being implemented than those that directly support the organization's strategic plan.

- An OPM mission statement from the Executive Team
Every Portfolio Development leader and his/her team must have a clear direction for themselves coming from a mission statement issued by the CEO or, even better, by the organization's entire Executive Team. It should be strategic, "action-oriented," and succinct. The OPM mission statement should directly support the mission statement for the organization. Here are several examples from our experience of successful organization's mission statements:
  - Example #1 (Fortune 500, For-Profit, Public Business Organization): "A relentless drive to invent things that matter: innovations that build, power, move, and help cure the world. We make things that very few in the world can, but that everyone needs. This is a source of pride. To our employees and customers, it defines (us)."
  - Example #2 (National, Nonprofit Charitable Organization): "Provide compassionate care to those in need. Our network of generous donors, volunteers, and employees share a mission of preventing and relieving suffering, here at home and around the world."
  - Example #3 (Federal Government Organization): "Protect the public health in the United States by ensuring the safety, effectiveness, and security of human and veterinary drugs, biological products, and medical devices; ensuring the safety of foods, cosmetics, and radiation-emitting products; and regulating tobacco products."

- Example #4 (Local, Nonprofit Charitable Organization): "Work in partnership with pastors and community leaders of (the metropolitan area) by providing them with tools and resources to strengthen their capacities to address pain and poverty."
- Example #5 (Global Nongovernmental Organization): "Build a better world by strengthening and improving the (parent organization) through the engagement of people who share a global mindset and support international cooperation–global citizens."
- Example #6 (Medium-Sized, For-Profit, Business Organization): "Provide people around the world with all the good things milk has to offer, with products that play an important role in people's nutrition and well-being."

In the context of these six examples, the following is a typical mission statement for an Portfolio Development Team: "Do a fair and unbiased review of the proposed and presently active projects/programs to define the ones that will have the biggest impact upon the organization's performance and reputation considering the resources that are available and the long- and short-range impact these proposed projects/programs will have on the organization."

Note: The purpose of the Portfolio Development Team is not to recommend the acceptance of a project or program but to ensure that only those projects or programs that have a high potential of being successfully implemented and resulting in the maximum benefits to the organization are recommended to be included in the organization's active projects or programs.

- Portfolio Development leader credentials and characteristics
  The Portfolio Development leader must be a *very* special, strategically focused individual and should possess the following credentials:
  - Is highly respected internally throughout the organization.
  - Is an experienced "organizational change agent" or "facilitator of organizational change."
  - Is a strategic thinker with systems analysis or systems engineering experience.
  - Must have no preconceived biases and/or prejudice.
  - Must have the integrity to recommend rejection of popular projects and projects sponsored and/or suggested by high-level executives.
  - Must value the time and skills of other personnel.

- Must be able to meet commitments that he/she makes.
- Must be creative in his/her approach to collecting data and analyzing the value a project/program will have
- Is proactive and preventive as an effective problem identifier, but can be an effective problem solver, when necessary, as well.

Additional traits to look for in selecting this Portfolio Development leader include:

- Understands the organization's strategy
- Trusted leader
- Self-starter
- Good listener
- Excellent communicator
- Politically savvy
- Has a detailed knowledge of the organization's business
- Understands processes and process improvement
- Customer-focused
- Passionate
- Motivator
- An excellent planner
- Excellent negotiation skills

Most important of all, this individual must serve as the "single point-of-contact" for the Executive Team, take ownership for the development of the Organizational Portfolio, and be able to either execute, or delegate the execution of, the remaining six activities in Phase I. In most organizations, this individual will either come from the Project Management Office (if the organization has one), Human Relations, or Finance.

- Portfolio Development Team role and responsibilities

  If the organization has 10 or more projects, programs, and portfolios combined, the Portfolio Development leader will have to recruit and select one or more Portfolio Development Team members to assist her/him.

  As an integrated group, the Portfolio Development Team should be comprised of individuals who will work together in support of the Portfolio Development leader to maximize the portfolio's value to the organization, such that each project or program, which they recommend that the organization invest in, produces more value than any other way the resources could be invested. The typical Portfolio Development Team would be made up of individuals from all of the key organizations involved in the portfolio of projects and programs. Typically, it would be made up of representatives from R&D, the Project Management Office, Marketing and Sales, Finance, Human Relations, and Manufacturing Engineering. Like all teams, in order to be successful, it will need to go through the five stages of team development: forming, storming, norming, performing, and transforming.[3] Unfortunately, this takes time.

The Portfolio Development Team's role is to share ownership with the Portfolio Development leader for the development of the Organizational Portfolio by executing the remaining six activities presented in this chapter.

The Portfolio Development Team's responsibilities will be determined by the Portfolio Development leader based on the unique demands of the organization's culture and the portfolio itself. (See the next input below.)

- Portfolio Development Team member credentials and characteristics

  Portfolio Development Team members should possess many of the same characteristics as the Portfolio Development leader, but not the same amount of experience. They should possess as many as possible of the following credentials:

  - Is respected by the departments submitting business cases for inclusion in the portfolio.
  - Has the potential to become an "organizational change agent" or "facilitator of organizational change."
  - Is a "systems thinker" with some systems experience.
  - Must not be self-serving.
  - Is proactive and preventive as an effective problem identifier, but can be an effective problem solver, when necessary, as well.

  Additional traits to look for in selecting Portfolio Development Team members include:

  - Willing to learn about the organization's strategy and its strategic processes
  - Trusted worker
  - Self-starter
  - Good listener
  - Good communication skills
  - Politically savvy and has leadership potential
  - Has a detailed knowledge of the organization
  - Coachable

- Customer-focused
- Energetic
- A capable planner and "toolsmith"
- Has a track record of being a successful project team member
- Excellent negotiation skills
- Must value the time and skills of other personnel
- Must be able to meet commitments that he/she makes
- Must be creative in his/her approach to collecting data and analyzing the value a project/program will have
- Has the ability to set aside personal interests for the good of the total organization
- Makes decisions based upon fact, not upon hearsay

Most important of all, these individuals must be willing to take shared ownership for the development of the Organizational Portfolio, and be able to support the Portfolio Development leader in carrying out the remaining six activities in Phase I.

Activities

- Talent recruitment and acquisition
  This is the practice of searching for the best potential candidates for the Portfolio Development leader. The Executive Team should have this responsibility as this individual needs to be well known and respected by every individual on the Executive Team. Typically, this is done at an Executive Team meeting where a list of potential candidates is developed and the positive and negative traits of each are discussed. Based on this discussion, the Executive Team should select the Portfolio Development leader.

  Because the members of the Portfolio Development Team have to possess special skills, it is best to delegate recruitment to the Human Resources Department and talent recruitment professionals to identify potential candidates who are then interviewed by the Portfolio Development leader.
- Candidate interviewing for the Portfolio Development leader assignment
  This is the practice of determining, amongst the top potential candidates for a given Portfolio Development leader position, which one *best* matches the position's requirements. This can be handled either via a telephone or face-to-face conversation, or both.

Even though this, too, is a specialty that is best led by the Human Resources and talent recruitment professionals, the CEO and/or COO need to be active participants in these meetings as they have a much better understanding of the difficulties that the individual candidates will face in preparing and managing the portfolio of projects. Organizations that have a Project Management Office will often give this assignment to the manager of the Project Management Office.

- Human Resource selection for the Portfolio Development leader assignment

  This is the practice of choosing the final candidate for the position and preparing to make the final offer. Even though the final offer is usually presented to the candidate by the Human Resource representative, the final selection of the candidate is the responsibility of the CEO or COO, taking into consideration any recommendations that are made by the Human Resource Department representative.

- Interviewing and selection of individuals for the Portfolio Development Team assignment

  The procedures for interviewing and selecting individuals to serve on the Portfolio Development Team are handled similar to the procedures used to interview and select the Portfolio Development leader, with the exception that the Portfolio Development leader is added to the individuals that do the interviewing and selection of the candidates for the Portfolio Development Team.

  Because the team member's assignment is usually a temporary short-term assignment (three to four weeks), candidates are usually selected from the internal operations based on their previous experience and knowledge of how the total organization functions. Typically, a number of the team members will be project managers who have already experienced managing projects/portfolios within the organization. It is good practice to have one member of this team hired from an outside source in order to give a nonprejudicial opinion related to which potential opportunities are recommended for implementation. This is often looked at as an additional unnecessary expense, but in the long run it is an excellent investment because it minimizes the risk of projects being selected based on internal politics rather than real value added to the organization.

Outputs

- A set of criteria aligned to the organization's mission to classify and rank the business cases
  Every Portfolio Development leader and his/her team must have a solid set of criteria that are aligned to the organization's mission identified as the first input for this activity to use as a "sounding board" for ranking the proposed business cases. (Note: Since this is also an input for Activity #2, it is covered in more detail below.)
- Portfolio Development leader selected and on board
  This is the practice of making the final offer to the candidate selected for the position, gaining acceptance of it, and ensuring that the newly hired (or transferred) person reports to his or her assigned manager.

  Even though this, too, is a specialty that is best led by the Human Resources professionals, the CEO and/or his/her delegate(s) are often active participants in producing this output.

  Once the individual is assigned as the Portfolio Development leader, he/she will review the requirements of the job and then he/she will jointly establish a detailed list of tasks and a timetable that is required to complete Activities #2 through #7 of Phase I.
- Portfolio Development Team member(s) are selected and on board (if appropriate)
  See the description for Portfolio Development leader selected and on board output above.

Activity #2: Classify the Business Cases Using a Qualitative, Quantitative, or Blended Model Based on the Potential Value

Once the Portfolio Development leader (and his/her team, if appropriate) has been selected, he/she must drive the completion of the remaining activities in Phase I in an objective and professional fashion using a set of agreed-upon classification criteria.

Here are the inputs, activities, and outputs for Activity #2:

- Input(s):
  - Portfolio Development leader
  - Portfolio Development Team members (if appropriate)
  - The business cases (proposed projects) to be classified

- Activities:
  - Business case validation
  - Document performance and project resource requirements for each project/program
  - Business cases that do not require additional resources
  - Select and use a set of criteria aligned to the organization's mission to classify and rank the business cases
  - Determine which classification model to use: qualitative, quantitative, or blended
- Output(s):
- An executive committee review and approved ranked-ordered list of projects and programs based on their potential value to the organization

Inputs

- Portfolio Development leader
- Portfolio Development Team members (if appropriate)
- The business cases to be classified

As part of a typical business cycle, each function should have submitted a set of business cases that they would like to start during the next business cycle. (Note: For details, see the two books mentioned above—*Value Proposition Development* and *Business Case Development*—in this Little Big Book series). On some occasions, the functional units submit project/program business cases for inclusion in the active approved activities within the organization between budget cycles. In these cases, the Portfolio Development Team handles them as a special case. These situations are discouraged, but in today's organization with a very fast-changing environment it is practically impossible to eliminate these special evaluation cases and still have the organization function effectively.

Based on our personal experiences during a budget cycle, a number of improvement opportunities are identified that have *not* gone through the Value Proposition Development stage or the Business Case Development stage either. Often it is *not* practical to ignore these improvement opportunities and as a result the Portfolio

Development leader will need to work with the individual functional area that is recommending or "nominating" these improvement opportunities to, at a minimum, prepare the data that are required for a business case so that the improvement opportunity can be fairly considered along with the other business cases. Often these last-minute improvement opportunities actually turn out to be "pet projects" sponsored by key executives within the organization and ignoring these key inputs could be politically "sensitive" and detract from the organization's potential performance. Unfortunately, these improvement opportunities have a tendency to increase the length of the budgeting cycle and, as a result, should be discouraged whenever possible.

The Portfolio Development leader and his/her Portfolio Development Team will focus their attention on classifying and ranking these business cases to develop the prioritized list of potential projects that will be considered to make up the approved portfolio.

Activities

- Business case validation
  The Portfolio Development leader should review each proposed project/program to ensure its business case is well-developed and includes practical and realistic estimates related to its goals, performance objectives, timing, and resource requirements. At a very minimum, realistic goals, performance objectives, timing, and resource requirements must be documented or the project/program will not be considered by the Portfolio Development Team for being included as an active project within the organization.
- Document performance and project resource requirements for each project/program
  We find that while the Portfolio Development Team is reviewing the individual business cases/value propositions, this is an excellent time to prepare a list of all the projects being evaluated and record what the projected impact is on the organization's performance and the resource consumption that is recorded in the business case or value proposition. We also suggest you record estimated implementation time and any risks that the group that prepared the business

case defined as impacting the project/program. This provides an effective bird's eye view of all of the proposed projects/programs being evaluated.

- Business cases that do not require additional resources
  Many of the business cases that are completed do *not* require additional resources for their implementation. Many of the identified improvement opportunities can be implemented within the normal activities that go on within the function and already in the approved budget; e.g., product engineering could have resources already budgeted to correct problems or make small changes to a current product. This would include activities like redesigning a part that is presently a steel machine part and replacing it with a plastic molded part. Another example would be when an operator suggested a different inexpensive tool be purchased to help them assemble a part and there's already money in the budget for miscellaneous tools. These improvement efforts require resources to evaluate and implement its activities that are normally part of the day-to-day job responsibilities of the individual organization. As a result, these business cases do *not* need to be considered as part of the organization's project portfolio. Only those business cases where the scope, magnitude, and impact fall outside of the normal job responsibilities of the function will be considered for inclusion in the organization's portfolio of projects and programs.

  Usually for these types of projects at the most, a value proposition is all that needs to be prepared and approved by the function's management team. In many cases, approval of these types of activities are either automatically approved by a memo or at a meeting where the activity is discussed and approved.

- Select and use a set of criteria aligned to the organization's mission to classify and rank the business cases
  The first step is to perform a general evaluation of each of the business cases to rate how it fits into the organization's overall business structure. While this step should have been done as part of preparing the proposed business case, we have found it helpful to double check it at this point in the portfolio development stage of the life cycle. Each of the business cases should be reviewed against the following nine items to determine if it is in minimum compliance with the related criteria (Table 2.1).

  Any business case that contains one of the six items that should meet the *"Must comply"* criteria, but falls short of it, must be dropped from

TABLE 2.1

Business Case Compliance with Key
Business Considerations

#	Item	Criteria
1	Mission Statement	Must comply
2	Policy	Must comply
3	Vision Statement	Should comply
4	Values	Must comply
5	Strategy	Must comply
6	Critical for Success Factors	Need not comply
7	Business Objectives	Must comply
8	Organizational Goals	Should comply
9	Strategic Business Plan	Must comply

further consideration for the organizational portfolio. Any business case that meets all six of the *"Must comply"* criteria items, but has an item that doesn't meet one of the two *"Should comply"* criteria, should still be given additional consideration during the evaluation cycle.

In addition, every business case should have, at a minimum, a "sustaining sponsor" identified and who will be held responsible for the success or failure of the project going forward.

• Determine which classification model to use: qualitative, quantitative, or blended

Now, there is a more granular and detailed approach to arriving at the criteria needed to classify and rank the projects and programs: one of three detailed classification or selection models: *Qualitative, Quantitative,* or *Blended*:

• A detailed selection model using *"Qualitative"* criteria provides an "anecdotal" or "subjective" perspective.
• A detailed selection model using *"Quantitative"* criteria provides an "empirical" or "objective" perspective.
• A detailed selection model using a *"Blended"* approach draws from both of the other two criteria providing a "dual" or "hybrid" perspective.

Depending on the set of criteria chosen to classify and rank the projects and programs as an input, you should determine which one of three detailed classification or selection models you are going to apply to arrive at that conclusion: a *Qualitative* one, a *Quantitative* one, or a *Blended* one.

Qualitative Classification Model[1]

There are five *Qualitative* models for classifying projects and programs.

1. *C-Level Executive's Pet Project*: This type of project/program is a "personal favorite" of a politically influential or powerful senior executive in the organization and is usually submitted "de facto" without going through the normal classifying or screening process. If this executive isn't the CEO, he needs to be sure that it gets ranked using the same model that is applied to *all* of the other projects/programs.
2. *The Operating Necessity*: This category is for those projects that are determined to be important in order to keep the organization running smoothly on a day-to-day basis.
3. *The Competitive Necessity*: This category is for those projects that are determined to be important in order to maintain the organization's competitive position in its market or industry.
4. *The Product/Service Line Extension*: This category is for those projects that are determined to be important based on the degree to which they are aligned with the organization's existing product or service line and extends it by filling a gap or strengthening a weakness to take it in a desired, strategic direction.
5. *A New Product/Service Line*: This category is for those projects that are determined to be important based on the degree to which they help execute the organization's strategic plan by adding something completely new to its line of products and/or services.

Quantitative Classification Model[2]

Quantitative models for classifying or prioritizing projects and programs are broken down into two subtypes: "**Profit/Profitability**" models, of which there are five approaches, and "**Scoring**" models, of which there are four approaches:

1. Profit/profitability models subtype
   • *Payback Period*: The initial fixed investment in the project is divided by the estimated annual cash inflows from the project. The resulting ratio is the number of years required for the project to repay its initial fixed investment. (Note: In order to qualify as a "Must Do" project, this value should be less than three years and, even better, less than two years.)

- *Average Rate of Return*: This is the ratio of the average annual profit (either before or after taxes) to the initial or average investment in the project.

The problem with the above two "simple" approaches is that neither one takes the concept of "the time-value of money" into account. Therefore, they should be considered *only* if interest rates and the rate of inflation are extremely low.

- *Discounted Cash Flow (DCF aka Net Present Value or NPV)*: This classification approach determines the net present value of all cash flows in the initiative by discounting them by the required "rate of return" (aka "hurdle rate" or "cut-off rate"). A positive net present value is desirable and preferable over a negative one.
- *Internal Rate of Return (IRR)*: The discounted rate ("k") that equates the present values of both the expected cash inflows and outflows as the result of undertaking the project or program. The value of "k" is found by extrapolation or "trial and error."
- *Profitability Index (aka Benefit–Cost Ratio)*: The present value of all future expected cash flows divided by the initial cash investment in the project/program. A ratio of >1.0 is preferable and the higher the better.

Figure 2.2 is a snapshot of an Excel® spreadsheet template that incorporates all five of the above "Profit/Profitability" quantitative models with easy-to-use formulas embedded within it.

To use the spreadsheet in Figure 2.2, simply change one or more of the inputs, such as the Discount (interest) rate, the annual costs, and/or the annual benefits. The formulas are already entered into the Excel file used to create this template. Be sure to double check the formulas based on the inputs to calculate the ROI, NPV, and Payback Period. Since the "Payback arrow" currently appears in the first year in which a Positive Value exists (Year #2) for the current "Cumulative benefits–costs" (Row #16), the arrow should be moved (manually) to reflect the year that is the result for *your* data. Everything else gets calculated for you.

2. Scoring models subtype include:
   - *Unweighted 0-1 Factor Model*: One or more OPT members are selected as "Classifiers" or "Raters" who score each project/ program using a set of relevant unweighted binary factors or criteria (either "Yes" or "No"). Typical scoring factors include: "Is a 'Green' Project," "Has High Potential Market Size ($)," "Has High

	A	B	C	D	E	F
1	**Financial Analysis for "*Project Name*"**					
2	**Created by:**		**Date:**			
3	Discount rate	5%				
4						
5	(Assume the project is completed in Year 0)			Year		
6		0	1	2	3	Total
7	*Costs*	100	100	100	100	
8	Discount factor	1.00	0.95	0.91	0.86	
9	**Discounted costs**	**100**	**95**	**91**	**86**	**372**
10						
11	*Benefits*	0	200	200	200	
12	Discount factor	1.00	0.95	0.91	0.86	
13	**Discounted benefits**	**0**	**190**	**181**	**173**	**545**
14						
15	Discounted benefits - costs	(100)	95	91	86	172
16	Cumulative benefits - costs	(100)	(5)	86	172	
17						NPV
18	ROI	46%				
19		Payback before Year X				

FIGURE 2.2
A spreadsheet with all five "profit/profitability" quantitative models.

Potential Market Share (percentage)," "Won't Require New Facilities," "Won't Require New Expertise," "Won't Sacrifice Quality," "Meets Profitability Target," etc. Then, the projects and programs are rank-ordered based on the relative number of "Yes" scores. (Note: While this model allows for multiple criteria to be used, it's assumed that all criteria are of equal importance with no scale. Quite often, this is *not* realistic.)

- *Unweighted Scaled Factor Scoring Model*: Conducted the same way as the previous model except that a 3-, 5-, 7-, or 10-point scale (with "1" as the lowest) is used and a discrete number is assigned to each unweighted factor or criteria. After that, the scores are added up, and the projects and programs are rank-ordered based on the relative score totals (with the highest score being classified highest, then, in descending order).
- *Weighted Factor Scoring Model*: One or more Portfolio Development Team members are selected as "classifiers" or "raters" who score each project/program using a set of relevant weighted and scaled values for each factor or criterion, based on a scale of 1 to 100 each. The weight (e.g., 0.1, 0.25, 0.5, etc.) of each criterion can be interpreted as the "percentage of the total weight

	A	B	C	D	E	F
1	**Weighted Factor Scoring Model for "*Project Name*"**					
2	**Created by:**			**Date:**		
3	**Criteria**	**Weight**	**Project 1**	**Project 2**	**Project 3**	**Project 4**
4	A	25%	90	90	50	20
5	B	15%	70	90	50	20
6	C	15%	50	90	50	20
7	D	10%	25	90	50	70
8	E	5%	20	20	50	90
9	F	20%	50	70	50	50
10	G	10%	20	50	50	90
11	**Weighted Project Scores**	**100%**	**56**	**78.5**	**50**	**41.5**

FIGURE 2.3
A typical "weighted factor scoring" quantitative model.

accorded to that particular criterion." (Note: Therefore, the total of all weighted factors *should* be 1.0 or 100 percent.) One of the potential weaknesses of this approach is the inclusion of one or more *marginal* criteria along with the *substantial* ones.

- *Constrained Weighted Factor Scoring Model*: This is conducted almost exactly as the previous model *except* that you define the marginal criteria as "*constraints*" instead of "*weighted factors.*" A constraint is a project characteristic that must be either present or absent in order for the project to be acceptable.

Figure 2.3 is a tabular example of a "*Weighted Factor Scoring Model*" in which Project 2 would be the *highest* ranked initiative (based on a scale of 1–100 points for each criteria based on its degree of compliance with it for each project) followed by Project 1, Project 3, and Project 4, respectively.

Figure 2.4 below is a graphic example illustrating the same data in ascending order by project number.

Blended Classification Model

A Blended Model for classifying or prioritizing projects and programs takes elements of both Qualitative and Quantitative models and combines them in a way that makes it easier to compare the various classified projects.

One of our favorite Blended Models for fulfilling this Phase I Activity is the Risk-Benefit Comparative Matrix Model, which presents all of the projects and programs being considered on a single graphic chart similar to the one for 25 projects in Figure 2.5.

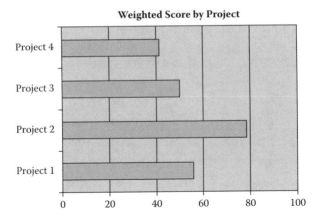

FIGURE 2.4
A typical "weighted factor scoring" graphic output.

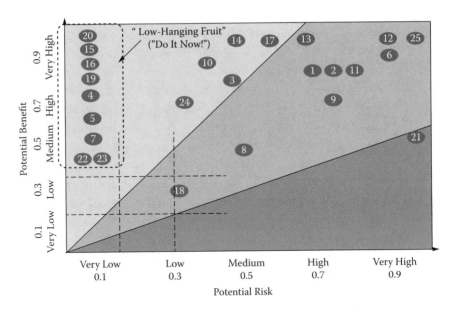

FIGURE 2.5
A typical "risk-benefit comparative matrix model."

In this example, we have classified the projects into five groups:

1. The "Low-Hanging Fruit" group (aka "Do-It-Now!") of nine projects (#22, 23, 7, 5, 4, 19, 16, 15, 20) in the narrow, dotted rectangle on the far left of the light-grey "Top Zone" that requires taking a very low risk for a relatively high return or benefit.

2. Five projects (#3, 10, 14, 17, and 24) that are in the "Very Desirable" group ("Must Do") in the center right of the same light-grey "Top Zone."

3. Eight more projects (#1, 2, 6, 9, 11, 12, 13, and 25) that are in the "Desirable" group ("Need to Do") projects, which are in the "Upper Middle Zone."

4. Two other projects (#8 and 18) that are in the "Less Desirable" group ("Nice to Do") projects located in the "Lower Middle Zone."

5. Project #21 that is in the only current "Undesirable" ("Defer") project in the dark-grey "Bottom Zone" because there is only a medium potential benefit (value) in return for taking a very high risk (threat).

Outputs

- An executive committee review and approved ranked-ordered list of projects and programs based upon their potential value to the organization

  The color-coded Comparative Matrix in Figure 2.5 also could serve as the basis for creating output for Activity #2: a rank-ordered, color-coded Organizational Portfolio. The five buckets or categories of projects and programs—"Do-It-Now" (Highest Priority), "Must Do" (Top Priority), "Need to Do" (2nd Priority), "Nice to Do" (3rd Priority), and "Defer at This Time" (No Current Priority—may be considered again in the future), and assigning each project or program to one of these categories.

  Such an output could look like the rank-ordered spreadsheet in Figure 2.6. (We recommend that the items on the spreadsheet be color-coded if the organization is using computers to view the document and/or can print the spreadsheet in color as it makes it much more effective and useful to the reader.)

Note of Caution: Although the ranking approaches described above are relatively straightforward and structured to provide management with a prioritized list of projects/programs, they should not be blindly accepted in place of good judgment and common sense. While they do provide excellent guidelines, often there are other factors that may be as unsophisticated and informal as "gut feel," intuition, or a key stakeholder's extraordinary level of enthusiasm related to a specific project. We have seen very lucrative projects fail due to the "ho-hum attitude" of the project sponsor

Rank-Ordered Organizational Portfolio of Projects & Programs					
Proj/Prog #	Do-It-Now	Must Do	Need to Do	Nice to Do	Defer
20	Yes				
15	Yes				
16	Yes				
19	Yes				
4	Yes				
5	Yes				
7	Yes				
22	Yes				
23	Yes				
14		Yes			
10		Yes			
17		Yes			
3		Yes			
24		Yes			
13			Yes		
1			Yes		
2			Yes		
11			Yes		
9			Yes		
6			Yes		
12			Yes		
25			Yes		
8				Yes	
18				Yes	
21					Yes

FIGURE 2.6
A simple "rank-ordered comparative spreadsheet."

and/or the management team while other projects of much lesser value are very successful due to the enthusiasm and dedication of a few individuals within the organization.

The rank order of the proposed projects/programs is an extremely important output as it will have a big impact upon the future of most organizations. It is for this very reason that the rank-ordered list of projects/programs needs to be reviewed and approved by the executive committee. We find that frequently there are business reasons that have a major impact upon the way individual projects/programs are prioritized within the organization. For example, a market study may be a key input to making a decision related to a future strategic alliance. Another example is a commitment that has been made by the CEO to the Board of Directors related to a specific initiative. Based upon the executive committee review, the rank-order comparison spreadsheet should be modified to reflect their its inputs before it is finalized.

Activity #3: Prioritize the Projects and Programs Based on the Resources Available

Before any project or program should be undertaken, a minimum of the following three types of resources need to be available, or are scheduled to be available, when they are needed. The lack of having access to any one of these three types of resources will almost certainly cause the project(s) to fail.

- Adequate **money** needs to be available to finance the project/program.
- **People** with the proper skills need to be available to design, implement, and operate the project/program.
- **Facilities** need to be available, in keeping with the needs of the project/program.

Over the coming time period, resources are freed up because some of the existing projects are winding down or being completed. Completed projects often result in improved processes that require less manpower and financial support to produce the desired outputs. In some cases, production schedules are being decreased, freeing up both direct and indirect personnel. In other cases, productivity improvement, which often exceeds 10 percent per year, is freeing up personnel to be assigned to newly proposed projects.

During a typical budget cycle, projected sales and revenue generation estimates are prepared. At the same time, the Board of Directors establishes a minimum acceptable profit level that would be acceptable to the stockholders. Based on these two projections, a maximum operating budget can be established. The first obligation of the organization is to fund the currently approved project and programs. The remainder of the operating budget can be considered for the support of newly proposed projects and programs.

As a result, the first part of a normal budget cycle focuses on defining the amount of resources that are required to support the currently approved projects and programs. It also takes into account the impact that projects, which will be completed during the time period, will have on resource requirements. The result of this part of the budgeting cycle is the identification of a timeline with acceptable requirements for personnel, facilities, equipment, and financial resources required to support the currently approved projects and programs as they relate to each department within the organization. These requirements are then compared to the presently available personnel, facilities, and equipment to identify reassignable resources. The financial resources projected to support the current projects and programs are compared to the proposed maximum operating budget to identify what financial resources can be applied to the newly proposed projects/programs. Typically, the personnel comparisons will be done by HR. The facilities and equipment comparisons will be done by the Industrial Engineering Department, and the financial comparisons is the function of the Controllers.

Note: In small- to medium-sized organizations, this identification of present resources that can be reassigned to support new projects/programs is often done in a much less structured manner.

Below are the inputs, activities, and outputs for Activity #3:

- Input(s):
  - An executive committee approved, ranked-ordered list of projects and programs based upon their potential value to the organization
  - Manpower plans organized by function/department
  - Documented performance and projected resource requirements for each project/program
- Activities:
  - Determination of personnel, financial, and facilities requirements
  - Assigning resources to prioritized projects/programs
  - Analysis and comparison of the value content for currently active projects/programs
  - Consideration of other resource alignment factors
  - Presentation to the Executive Team
- Output(s):
  - A list of new, resource-leveled, rank-ordered projects/programs that will be included in the organization's active projects
  - A list of new, rank-ordered projects/programs with no resources assigned and are deferred

- A list of projects/programs that are *not* recommended for implementation and, therefore, are dropped
- A written description for all approved projects and projects put on hold that includes the goals, objectives, and resource requirements
- Documented performance and projected resource requirements for each project/program

Inputs

- An executive committee-approved, ranked-ordered list of projects and programs based on their potential value to the organization
  This is the sole output from the previous Activity (#2) in which each project and program has been evaluated, prioritized, and placed into one of the five buckets or categories of projects and programs below:
  1. *"Do-It-Now"* (Highest priority; may be "low-hanging" fruit that can be done now with minimal effort and consumption of limited resources)
  2. *"Must Do"* (Next highest priority)
  3. *"Need to Do"* (3rd priority)
  4. *"Nice to Do"* (4th priority)
  5. *"Defer at This Time"* (No current priority; might be considered again in the future)
  These prioritized projects and programs can be displayed in a graphic format (Figure 2.5), simple tabular format (Figure 2.6), or a spreadsheet format where they can be further defined in the ensuing steps in the process.
- Manpower plans by function/department
  Once these projects and programs have been categorized and listed in rank order, the various functions or departments in the organization can perform "Manpower Planning" on each of them, in order of priority.
  Manpower Planning consists of forecasting the right number and kinds of people at the right place, at the right time, doing the right things for which they are suited for the completion of the projects and programs in the organization's portfolio. The result will be Manpower Plans by function/department based on those forecasts. These plans will be needed along with the rank-ordered Organizational Portfolio in order to complete this activity.

- Documented performance and projected resource requirements for each project/program
  All major programs will have realistic business plans prepared and documented for them that include their goals, performance objectives, timing, and resource requirements.

Activities

- Determination of personnel, financial, and facilities requirements
  Requirements for every project/program need to be addressed in three different ways:
  1. Personnel requirements
  2. Financial requirements
  3. Facilities requirements
  The first consideration is personnel requirements. In many organizations, the primary focus is on controlling the number of employees within the organization. This focus is brought about because management realizes that it makes a major commitment to an individual once he or she is hired. These commitments include the individual's salary, benefits, and continued employment. In these organizations, the employee turnover rates are extremely important to keep at a minimum level because hiring and training employees is a major financial expenditure.
  These organizations do everything possible to provide meaningful, value-added employment for the current employees and only add staff when there is a long-term potential for new staff to be added. These organizations look at their present employees as part of their assets rather than an expense. These types of organizations go out of their way to prepare careers for the current employees and rely on hiring supplemental temporary staff, consultants, and/or outsourcing to handle any short-term requirements.
  In these cases, short-term requirements become more a matter of having sufficient funds available rather than personnel available. In many cases, having personnel available does not satisfy the needs of the proposed projects/programs because the people that are available do not have or are not capable of learning the required skills in time to support the proposed project/programs.
  In the cases where the number of employees needs to be controlled, the number allocated to a specific location is usually defined by the

Executive Team. The number of employees presently employed is usually maintained by either the HR Department or the Industrial Engineering Department. The workload for the present employees is defined during the budgeting cycle. Surplus employees are identified as a result of decreases in projected workload, committed productivity improvements, and continuous improvement programs.

Ideally, these surpluses are defined by the various functions within the organization and the skills that the surplus employees presently have. Based on the individual organization's job descriptions, this type of information can be supplied by the organization responsible for coordinating the preparation of the budget in conjunction with the HR Department. This information is usually broken down by surplus per department per month.

The second consideration is the financial requirements. In organizations where the primary focus is on financial controls rather than human resources, the approach to determining which projects/programs will be included in the approved portfolio of projects/programs is much simpler. In these cases, where employees are looked at as a cost rather than an asset, surplus employees are assigned to new programs/projects only if that is the most economical way of doing business.

In this case, the decisions are made based on the cost of hiring temporary employees, training present employees in the skills required for the new assignments, the use of consultants, and the cost of hiring new full-time employees. In these cases, the Executive Team, in conjunction with the Board of Directors, establishes a targeted maximum yearly expenditure for the organization. This money is first distributed to the current ongoing projects and the remainder is set aside for discretionary spending, which includes new projects/programs. Today's actual expenditures plus a good understanding of projected increases and decreases in activities and the projected committed performance improvement results would typically provide Finance the information needed to project financial requirements for the current activities. The delta between these financial projections and the maximum expenditures set by the Executive Team defines the discretionary spending that is allotted for activities like new projects/programs.

The third consideration is the facilities requirements. The two major facilities variables that need to be considered here are equipment and space.

Equipment facilities primarily refers to major items like mainframe computers, test equipment, milling machines, punch presses, clean room facilities, automobiles, and major pieces of software (CRM, MRP, etc.). Often as current projects become obsolete and/or eliminated, equipment facilities are freed up for the new projects/programs. When this is not the case, the additional costs need to be factored into the financial requirements along with the personnel effort required to install them.

Space is the other facilities variable that needs to be considered. Space includes warehousing, manufacturing floor space, office cubicles, filing areas, reproduction areas, and conference room space. Space is an important consideration because additional space to support new programs/projects would typically be established while the current products/programs are still utilizing their current space. Often temporary setups need to be established for the new programs/projects until activities on the current projects are reduced or eliminated. The organization's Industrial Engineering Department (if one exists) should be able to provide the Portfolio Development Team with the organization's available facilities.

• Assigning resources to prioritized projects/programs
As the Portfolio Development Team starts the job of assigning resources to the prioritized projects/programs, it has five important inputs that will be used to guide their activities:
1. A list of rank-ordered projects/programs.
2. A list of surplus employees by skills and department. That may include the maximum number of employees allotted for the organization.
3. The maximum amount of discretionary spending money that can be allotted to the projects and programs.
4. An individual from a function/department like industrial engineering that understands the availability of facility resources.
5. The organization's strategic plan.
Starting with the highest priority project/program, the Portfolio Development Team will analyze and define the resource requirements for the project/program and assign personnel, financial, and facility resources to each of the programs/projects. Each time a resource is assigned, the remaining amount of resources available to lower priority projects is decreased by the value of the assigned resource. This process is repeated until the available resources are

exhausted. Unfortunately, this usually occurs before all of the proposed programs'/projects' resource requirements have been fulfilled. When this occurs, the Portfolio Development Team needs to look at currently active projects/programs in an attempt to maximize the value added to the organization in the total portfolio of projects and programs.

- Analysis and comparison of the content value of the current active projects/programs

  Very often, identified available resources are totally used up before all of the proposed new projects have resources assigned to them. In these cases, an assessment should be made of the current approved projects/programs to ensure that their value-added content and projected impact based upon current data would justify them being canceled in favor of one or more of the proposed new projects/programs. Often, in today's rapidly changing business environment, projects that were projected to be highly lucrative have diminished in luster and it is more beneficial to the organization to terminate them in favor of a newly proposed project. When this occurs, the Portfolio Development Team should make a list of the current projects/programs that they recommend should have the resources reduced or should be eliminated, and document how these work resources would provide better value to the organization if they were applied to specific proposed projects/programs.

- Consideration of other resource alignment factors

  As the reader must realize by now, this activity is very complex. However, there are two other factors that must be considered when assigning resources: timing and possible future shock.

  1. Timing: For example, consider the following. The highest priority project is scheduled to start May 1 and requires three programmers to be assigned for the rest of the year. A lower priority project is scheduled to start on January 1 and requires three programmers until June 15. The data indicates that there were four programmers available for the entire year. In this case, the simplest answer would be to assign three of the programmers to the highest priority project starting May 1 and hire temporary supplementary programmers to handle the needs of the other projects. In that case, we would have three internal programmers who would not be utilized for four months. To eliminate this waste of the programming resources, the Portfolio Development

Team would need to look into the possibility of using four programmers on the second project so that it could be completed by May 1 and/or the possibility of slipping the programming needs in the high level project by a month and a half.

2. Possible Future Shock: Assigning too much change activity to an individual part of the organization might cause it to go into future shock. An individual part of the organization can be involved in so many different change initiatives that it is no longer able to handle the stress that is put on the area as these changes are taking place. This situation is called *future shock* and, at that point, the area becomes dysfunctional. To minimize the possibility of an individual organization going into future shock, the Portfolio Development Team needs to understand the amount of change activity that is going on within the individual areas of the organization and recommend to the executive committee that projects are scheduled so that the stress within individual parts of the organization does not exceed its ability to function in a responsible manner. As a general guideline, no more than two major changes should be going on in an individual part of the organization simultaneously. That is the maximum that the area can handle effectively.

- Presentation to the Executive Team

  The Portfolio Development Team should schedule a meeting with the Executive Team to present and review its recommendations. This presentation should include a detailed analysis of how available resources were assigned to the project/programs based on their prioritized order. It also should present the current projects that the Portfolio Development Team is recommending to have their activities reduced or to be completely eliminated and how these actions would free up resources that could be applied to the newly proposed projects/programs.

  The Portfolio Development Team should present as well the list of projects/programs that did not have resources assigned to them along with recommendations related to their future potential or if they should be dropped from consideration. Sometimes this results in the Executive Committee increasing the resources available in order to include additional proposed programs/projects in the active portfolio.

Outputs

As a result of Activity #3, the proposed lists of new projects/programs are divided into three major output lists:

1. A list of new, resource-leveled, rank-ordered projects/programs that will be included in the organization's active projects.
2. A list of new, rank-ordered projects/programs with no resources assigned and are deferred.
3. A list of projects/programs that are *not* recommended for implementation and, therefore, are dropped.

Also,

- Documented performance and projected resource requirements for each project/program.
- A written description for all approved projects and projects put on hold that includes the goals, objectives, and resource requirements.

In addition, a list of current active projects/programs that are recommended for completion and another list of those current active projects/programs that are recommended to have their resources reduced or termination in favor of one or more of the newly proposed projects/programs should be prepared and presented by the Portfolio Development Team.

Activity #4: Select Projects and Programs for the Portfolios and Assign a Portfolio Leader to Oversee Them

The purpose of this activity is to group together projects/programs in a manner where a Portfolio leader can be assigned to monitor them with the objective of keeping them on time and within budget while producing the desired business results. Below are the inputs, activities, and outputs for Activity #4:

- Input(s):
  - A list of all resource-leveled, rank-ordered projects/programs
  - A list of all current active projects/programs/portfolios
  - A list of qualified portfolio leaders
  - A list of initiating sponsors for all current and approved projects/programs

- Activities:
  - Select the subportfolio(s) for the current Organizational Portfolio
  - Match each subportfolio by assigning a qualified Portfolio leader to it
- Outputs:
  - A list of resource-leveled, rank-ordered proposed projects/programs in the portfolio with each one's assigned portfolio leader
  - A list of initiating sponsors for all current and approved projects/programs
  - A list of portfolio leaders with subportfolios and projects/programs assigned
  - A list of all current active projects/programs/portfolios

Inputs

- A list of all resource-leveled, rank-ordered projects/programs
  To accomplish this activity, the list of current, active projects/programs and the list of prioritized business cases that are recommended for having resources assigned to them are combined. This allows similar or relevant projects/programs to be grouped together into separate portfolios.
- A list of all current active projects/programs/portfolios
  The number of projects/programs that can be effectively managed by a Portfolio leader will vary based on the complexity of the respective project/program, the core competencies required to manage a specific type of project/program, the physical location at which the project is being implemented, and the timing of the project/program. As a general rule of thumb, an effective Portfolio leader will be able to handle from three to five projects simultaneously. Very often, individual portfolios of projects/programs are created for new products, marketing, information technology, and personnel-related initiatives.
- A list of qualified Portfolio leaders
  It is extremely important that knowledgeable and effective individuals are recruited, selected, and assigned as Portfolio leaders. To assist the Human Resources Department in finding and selecting competent Portfolio leaders, a comprehensive job description should be created that is modeled after the one provided below for a Data and Coordination Center.

DCC PORTFOLIO LEADER

Background context: The Data and Coordination Center (DCC) is responsible for coordination of and data capture for a multicenter consortium of Clinical Centers of Excellence providing annual screening examinations, mental and physical health treatment, public health reporting, and investigation of health outcomes among a heterogeneous, multilingual population of over 34,000 patients. The Portfolio leader will have responsibility for managing the DCC's Portfolio of Projects.

Position summary: The Portfolio leader's responsibilities will be to oversee and coordinate three to five complex performance improvement, data management, and scientific investigation projects to ensure that multiproject deliverables and timelines involving a diverse team of physicians, epidemiologists, biostatisticians, social workers, informaticists, and data management specialists are identified, scheduled, and met. The Portfolio leader will report directly to the DCC principal investigator/director.

Primary Responsibilities:

- Direct and support the development of a portfolio of comprehensive project plans and timelines in collaboration with the four DCC Core Teams (Health Outcomes, Clinical Coordination, Community Outreach/Social Services, and Data Management) to implement strategic priorities; establish goals, define deliverables/timeframes, and evaluate performance with the goal of ensuring that federally mandated deliverables are completed and provided on a timely basis.
- Direct and support tracking, monitoring, and controlling the above-described portfolio of comprehensive project plans and timelines with the goal of ensuring that federally mandated deliverables are completed and provided on a timely basis.
- Has full accountability for the creation, implementation, and facilitation of a portfolio of projects related to federal grant program deliverables including public health reporting and disease surveillance, data entry and cleaning, creation of analytic datasets, and coordination of clinical outcomes reporting, including

administration of scientific writing groups and provision of assistance in manuscript preparation and progress tracking.

- Facilitate timely and responsive integration amongst and between the Health Outcomes, Data Management, Clinical Coordination, Social Services/Community Outreach, and Informatics teams, principal investigators, the funder, and other key stakeholders, including responder organizations, public health, and labor unions.
- Post on the shared drive and maintain a read-only portfolio comprised of project-related records to allow for secure access to multiple, cross-organizational project plans for managing internal and external resources to ensure all schedules and deliverables are met.
- Develop and maintain a portfolio status reporting system to track implementation schedules, deliverables, and resources on a weekly basis.
- Develop and maintain processes to identify and report obstacles/barriers that could prevent the DCC from meeting its portfolio and project objectives.
- Act as liaison between the Leadership Core members and Core Team leaders.
- Responsible for internal cross-project communications. Ensure that all team members are informed of changes in direction or any other information related to their ability to perform their role on one of the projects or the program as a whole.
- Take the lead role in issue resolution for projects, mediate conflicts, and ensure that key people are included in issue resolution on a timely basis.
- Work with the Institutional Review Board and the HIPAA Compliance Team to ensure conformance with regulatory requirements.
- Identify process and workflow "bottlenecks" and recommend changes required to implement the DCC's strategic goals.
- Support the creation of project plans with milestones and work with the project managers and their team members to assure that project deliverables are completed on time and within budget.

- Track, communicate, and manage all portfolio implementation issues.
- Review all portfolio and project reports for accuracy and compliance with programmatic goals.

Experience/Requirements:

- 10+ years of project management experience with at least 2 of those years in a leadership role within a Project Portfolio environment.
- Experience in either medical informatics, public health, biopharmaceutical, biotechnology, medical devices, or a Clinical Research Organization (CRO) or Data Center is highly desirable.
- Understanding of data management for public health and scientific reporting is essential.
- Must have excellent interpersonal and leadership skills, as well as strong communication, problem solving, and organizational skills.
- Experience in planning and implementing biomedical, operational, and/or administrative information systems applications and integration strategies is strongly desired.
- Excellent verbal and writing skills and demonstrated aptitude for problem solving.
- Excellent analytical skills.
- Must possess working knowledge and understanding of relational databases.
- Must be proficient in accounting principles as they relate to the design of performance reports.

Education/Certification:

- Master's degree in medical informatics, epidemiology, biostatistics, information technology, computer sciences, healthcare management/administration, public health, or its equivalent required; doctoral degree preferred.
- Certification in Project, Program, and/or Portfolio Management, Lean Six Sigma, or their equivalent is required.

In sum, just like the Portfolio Development leader, the Portfolio leader must be a *very* special individual. However, the Portfolio Development leader is a temporary type operation and often this individual's permanent job will be in strategic planning, personnel, or managing a project office. In large organizations where there are a number of Portfolio leaders and/or project managers, a department called *Project Management Office* (PMO) is formed. In these cases, the Portfolio leaders and the professional project managers reside as members of the PMO and they are responsible to the manager of the PMO, who is responsible for coordinating the activities that are going on between all of the Portfolio leaders.

Each Portfolio leader is only responsible for managing a delegated set of projects/programs. That being the case, the Portfolio leader should focus on planning, executing, monitoring, and controlling a set of high-priority projects and programs. He/she should have the following credentials:

- Possesses at least a bachelor's degree, but, in some cases, a master's degree
- Is highly respected throughout the organization
- Possesses certification in Project Management or Program Management from a recognized professional society, or hold a master's degree in Project Management from an accredited institution.

Additional traits to look for in selecting the Portfolio leader include:

- Understands the importance of his/her portfolio of projects and programs
- Trusted leader
- Self-starter
- Good listener
- Excellent communicator
- Politically savvy
- Has detailed knowledge of the business
- Understands processes
- Customer-focused
- Passionate
- Strong motivator
- Driven by the completion of deliverables

- Has a track record of getting projects done on schedule and within budget
- Excellent follow-up skills

Most important of all, this individual must be able to serve as the "single point-of-contact" for his/her portfolio of projects and programs and take ownership for their execution and completion.

- A list of initiating sponsors for all current and approved projects/programs

Activities

- Select the subportfolio(s) for the current Organizational Portfolio

Because the number of projects/programs that can be effectively managed in a subportfolio by a Portfolio leader will vary ("Rule of Thumb" = 3–5 per Portfolio leader is preferable) based upon the factors identified in the inputs section above, a decision must be made as to the number of projects/programs that should be active at any given time. However, this decision also will have to include the number of qualified Portfolio leaders that are available.

Note: All the active projects/programs make up the Organizational Portfolio. Because the individual Portfolio leader can only effectively manage a limited number of projects/programs, the Organizational Portfolio often is divided into subportfolios that are managed by individual Portfolio leaders. These subportfolios are usually grouped together in some logical fashion to facilitate the effectiveness that the assigned Portfolio leader can manage. Very often these subportfolios are identified based on projects/programs that are going on within an individual function. Typical examples would be portfolios related to marketing, new product development, procurement, or construction, or, for example, if there are a total of 24 projects/programs in the currently active Organizational Portfolio in six different groups or subportfolios, you will need six qualified Portfolio leaders to adequately manage the workload. If there are currently only five qualified Portfolio leaders, then the Portfolio Development leader must determine which subportfolio will have to be deferred until another qualified Portfolio leader can be recruited. Often the grouping of these subportfolios is based on the competencies of the Portfolio leaders who are available.

- Match each subportfolio by assigning a qualified Portfolio leader for it Once the specific subportfolios have been selected, the Portfolio Development leader must match each of them with a qualified Portfolio leader to bring them through to successful completion. This "matching" should consider the type of application area know-how, amount of experience, and level of maturity required to appropriately address the challenges and opportunities that are unique to each subportfolio.

Outputs

- A list of resource-leveled, rank-ordered proposed projects/programs in the portfolio with each one's assigned Portfolio leader
- A list of initiating sponsors for all current and approved projects/programs
- A list of Portfolio leaders with subportfolios and projects/programs assigned (Figure 2.7)
- A list of all current active projects/programs/portfolios

Activity #5: Identify a Sponsor and Project Manager for Each Project and Program

Below are the inputs, activities, and outputs for Activity #5:

- Input(s):
  - A list of qualified project managers
  - A list of resource-leveled, rank-ordered portfolios and projects/programs with portfolio leaders identified
  - A list of initiating sponsors for all current and approved projects/programs
  - Documented performance and projected resource requirements for each project/program
- Activities:
  - Obtain a qualified sustaining sponsor for each project/program
  - Review each project/program to identify which ones do not require a project manager
  - Match each recommended project or program to a qualified project manager that has the required skills

Portfolio Project Manager	Project Sub-Portfolio	Project/ Program #	Project Rank
Carlos	A	20	1
		23	9
		14	10
		13	15
Jacquie	B	15	2
		7	7
		17	12
		6	20
Frank	C	16	3
		10	11
		1	16
		11	18
Paula	D	19	4
		22	8
		12	21
Tammi	E	4	5
		3	13
		2	17
		25	22
Dave	F	5	6
		24	14
		9	19

FIGURE 2.7
A list of Portfolio leaders with the subportfolios assigned.

- Output(s):
  - A list of projects and programs that require project management with the name of the project manager responsible for the project/program
  - A list of the project managers with the name of the projects that are assigned to each
  - A list of projects that do not require a project manager to be assigned to them

Inputs

- A list of qualified project managers
  The HR department should be able to scan the personnel records and identify a list of individuals who have the qualifications to serve as a project manager. If they do not have this capability, the Portfolio Development Team can review past projects to identify potential qualified project managers. If that does not provide adequate resources, the organization can subcontract the assignment to already certified project managers or hire new employees that are already certified as project managers.
- A list of resource-leveled, rank ordered portfolios and projects/programs with portfolio leaders identified
- A list of initiating sponsors for all current and approved projects/programs
  All projects/programs are required to have an initiating sponsor for them to be accepted as an active project within the organization. Once the project is approved, it should have a sustaining sponsor assigned as well. In some cases, the initiating sponsor and the sustaining sponsor are the same person. In other cases, the initiating sponsor may not have the time to become deeply involved in the specific project and the sustaining sponsor must be assigned to maintain contact with the project:
  - *Initiating sponsor* is the individual or group with the power to initiate or to legitimize the change for all of the affected people in the organization.
  - *Sustaining sponsor* is the individual or group with a political, logistic, and economic proximity to the people who actually have implemented and/or who have been affected by the project/program.
    Often we talk about initiating sponsors as senior management, and sustaining sponsors as middle management, but that's not always necessarily the case. Often these sponsors can be someone in the organization who isn't in the direct management reporting line, but who has significant power of influence because of relationships with the people affected by the change, past success, knowledge, or power.
- Documented performance and projected resource requirements for each project/program

Activities

- Obtain a qualified sustaining sponsor for each project/program
  Initializing sponsors are normally the chief officer in charge of a function or an executive officer in the organization. Although every business case is required to have a sustaining sponsor, the sustaining sponsor is often at a lower level than the initiating sponsor or in a different part of the organization than the initiating sponsor. The sustaining sponsor is the primary representative of the management team that interfaces with the Portfolio Development Team. He/she maintains a close relationship and understanding of the project that he/she has accepted as a management representative.
  To gain the support of the appropriate initializing sponsor, an updating meeting should be scheduled for each of the proposed new projects that are recommended for implementation. The project manager or a member of the Portfolio Development Team should attend this meeting along with the individual who prepared the business case. The purpose of this meeting is to get the individual executive to agree to be the sustaining sponsor of the project. Failure to get a sustaining sponsor of the project often causes the project to be put on hold.
- Review each project/program to identify which ones do not require a project manager
  The complete list of approved projects/programs should be reviewed to identify the ones that do not require a project manager's support to minimize the risk of the project/program failing to meet requirements. In matrix organizations, where responsibility for project management is frequently shared between functional work groups and the Program Management Office (PMO), short-term, small group, low risk type projects can be managed effectively by the functional Project Team manager of the group that is assigned to do the work of the project. The Portfolio Development Team should prepare a list of the approved projects that do not need to have a project manager assigned to them.
- Match each recommended project or program to a qualified project manager
  Project managers need to be trained and understand the project management process in order to minimize the organization's risk of failed projects. We recommend that only people who have been trained and certified as a professional project manager be assigned to the project management role. Most Human Resource Departments

PROJECT MANAGER

Experience:

- Minimum five years business experience.
- Has worked with EPMO QA manager or business analyst to ensure implementation of PM standards, processes, and support services.
- Can resolve issues related to client relations, governmental relations, project quality, project risk, and project safety.
- Manage vendor relations and procurement related to the assigned projects.

Functional Competencies:

- *Business:* Business acumen, follow-up, negotiation, decision making, planning, and organizing
- *Leadership:* Driving execution, meeting leadership, and mobilizing resources
- *Personal:* Adaptability, managing conflict, building strategic work relationships, and communication
- *Technical:* Experienced PC skills and expertise using MS-Office® suite of applications, including MS Project and Visio

Skills and Abilities:

- Competency in project management processes, including planning tasks and allocating resources, risk management, organizational change management, issues management, time management, financial management, HR management, working in teams, quality management, monitoring and reporting, documentation and recordkeeping.
- Ability to plan and facilitate meetings.
- Strategic, conceptual analytical thinking and decision-making skills.
- Adaptability and flexibility including ability to manage deadline pressure, ambiguity, and change.

- Good conflict resolution and negotiation skills within a context of political sensitivity and conflicting interests.
- Good analytical skills and the ability to present findings in a clear, concise manner.
- Ability to advise on complex matters to nonspecialists; ability to communicate effectively with senior management.
- Good oral and written communication skills and exceptional interpersonal skills.
- Possesses the ability to work well with people from many different disciplines with varying degrees of technical experience; competence in clear, concise, and tactful communication with senior management, clients, peers, and staff.
- Budget management skills; ability to analyze and review financing plans and related budgetary information to determine the impact on a project is required.

Education/Certification:

- Bachelor's degree in information technology, computer sciences, business administration, or its equivalent required; master's degree preferred.
- Certification in Project Management, Lean Six Sigma, or its equivalent is required.

will have a job description for a qualified project manager. The following is a typical one.

Although we would recommend that the project manager's job be filled by an individual who has been certified as a project manager by a reputable professional association, it is not mandatory. It is a just a good way of reducing the size of the applicant pool and minimizing the risk related to having an unsuccessful project.

Having a detailed understanding of how to manage a project is just part of the consideration when we are assigning project managers to individual projects. Often it is equally important to have an excellent understanding of the technologies involved and the processes that are relevant to the project in order to have a successful project that is completed on time and within budget. It is not even necessary that the project management assignment be full time. We have seen many projects that are very successful where the project was managed part-time by an individual who had assignments within the project. For example, we have seen very successful new product development projects managed by individuals from product engineering who were deeply involved in the design of the product while he/she is also serving as the project manager.

Another important consideration is the workload that is assigned to the individual project managers. The number of projects that individual project managers can handle varies from two to five depending upon the complexity and timing of the projects.

Outputs

- A list of projects and programs that require project management with the name of the project manager responsible for the project/program
- A list of the project managers with the name of the projects that are assigned to each
- A list of projects that do not require a project manager to be assigned to them

Activity #6: Define High-Level Project Milestones and a Budget for Each Project and Program

- Input(s):
  - Business case for each approved project
  - Project managers, initiating sponsor, and sustaining sponsors assigned to each approved project/program
- Activities:
  - Project team manager assigned to each approved project/program
  - High-level milestone scheduling
  - High-level cost estimating and budgeting
  - Draft executive summaries for each proposed project/program
- Output(s):
  - High-level budget set for each project
  - Key management checkpoints are defined and documented
  - The project team is defined and assigned
  - Executive summaries are prepared for each proposed project/program

Inputs

- Business case for each approved project
  Each approved project/program should have a business case prepared for it before manpower is approved and it is included in the Organizational Portfolio. On occasion, projects/programs are approved without a comprehensive business case prepared for them. In these cases the Project Team manager involved in conducting Activity #6 needs to take the additional time required to generate the backup information needed to prepare a budget and a time schedule for the program/project.
- Project managers, initiating sponsors, and sustaining sponsors assigned to each approved project/program
  Before Activity # 6 is started, a project manager, initiating sponsor, and a sustaining sponsor need to be assigned and available to actively participate in this activity.

Activities

- Project team manager assigned to each approved project/program
  A common agreement should be reached among the project manager, initiating sponsor, and the sustaining sponsor related to the Project Team manager who will be responsible for being in charge of implementing the project. He/she is responsible for the day-to-day activities and coordination related to the project to which he/she is assigned. Typically, this individual serves as the Project Team manager. For most major projects, this is a full-time assignment.
- High-level milestone scheduling
  There is always a debate about what comes first: the schedule or the manpower and resource requirements.. We find that the best approach is to establish your major milestones and then determine if adequate resources are available to complete the required activities in keeping with the milestones. If not, then the milestones need to be adjusted.

  Considerations related to other organizational initiatives (especially those projects focused on achieving an incremental improvement) may be interrelated and, therefore, may have an implied implementation sequence. This has a direct effect on how related initiatives are prioritized. The success of a mission-critical initiative may be based on the availability of information or implementation of another initiative that is of lower importance. This relationship between initiatives may raise the priority of the mission-critical initiative. The following *three* situations in which initiative dependencies or interproject dependencies play an important role in prioritization include:

- *Sequential Dependencies*: Dependencies that rely on information or results produced from another initiative or provide information or results that are required by another initiative. For example, if a Sales Forecast initiative produces information or results for a Budgeting initiative, and the Budgeting initiative produces information or results required for a Financial Planning initiative, then all three are sequentially dependent.
- *Concurrent Dependencies*: Initiatives that rely on the same information or result from the same initiative prior to their execution. For example, if a Procurement initiative provides information or results needed to successfully implement both Accounting and Inventory initiatives, then both the Accounting and Inventory initiatives are considered to have a concurrent dependency upon the Procurement initiative.
- *Mutually Exclusive Dependencies*: Initiatives that require the analysis of the results of a previous initiative in order to determine which of the two initiatives needs to be implement next. For example, a Marketing initiative may provide information that allows a decision on whether a Cost Reduction initiative or a Customer Service initiative will have a greater impact on market share.

Once the Project Team manager is on board, a meeting to establish the high-level schedule for the project is called. The individuals participating in this meeting should include, but not be limited to, the Project Team manager, the sustaining sponsor, and the project manager. Each person that is attending this meeting should have previously become very familiar with the business case that supports the approved project. Typically, the business case will include some key dates and assumptions on which the project is based. Often it includes key dates that are important to adhere to in order to produce the value added that was projected in the business case.

One very effective way to develop a high-level schedule is to do a high-level block diagram of the activities that are required to implement the project. To use an example that many of you are very familiar

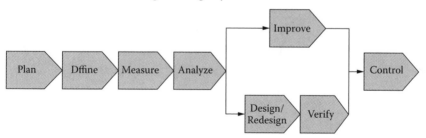

The Six Sigma Improvement and Design/Redesign Cycles Combined

FIGURE 2.8
The Six Sigma improvement and design/redesign cycles combined.

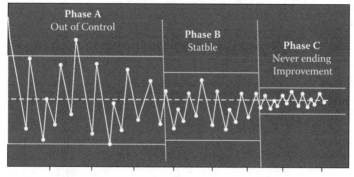

Reducing Veriation Objectives

FIGURE 2.9
A project objective to reduce variation.

with, we will look at scheduling a Six Sigma project. Figure 2.8 is a block diagram of the Six Sigma implementation process.

By analyzing the business case, they were able to determine that the Six Sigma improvement was focused on reducing variation, not on setting new standards of performance (Figure 2.9).

Based on this, they are able to define that the process would be made up of five major phases:

1. Define
2. Measure
3. Analyze
4. Improve
5. Control

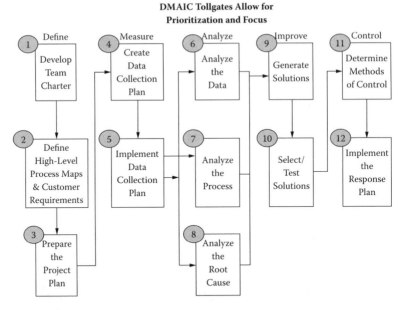

FIGURE 2.10
Six Sigma tollgates for reduced variation objective project.

They then looked at each of the five phases to define what major activities took place. During these phases, each of these major activities was controlled by installing a tollgate at the end of the activity that would be used to ensure the activity was done, measure how well it was done, and to see if the project was ready to move on to the next major tollgate (Figure 2.10).

In this particular case, 12 tollgates were defined. The business case defined that the project would last for 90 days, but did not define the starter in point for the project. As a result, the duration was defined but the timing was not.

They next focused on defining the areas within the organization that would be impacted when the project is implemented. This resulted in defining five departments that would be impacted. They then looked at each of these five departments to determine what other change activities and stresses are being applied to determine if the additional stress of this change would put any of the departments into Future Shock. Future Shock is defined as the point at which people can no longer assimilate change without displaying dysfunctional behaviors. (See H. James Harrington's book, *Change Management Excellence: The Art of Excelling in Change Management*

(Paton Press, 2006). Because in this case the project was not dependent upon any inputs other than human resources and there were no other major project activities going on in the impacted areas, the decision was made to start the project as soon as practical.

The next thing they did was to define the makeup of the Project Team that would be responsible for designing and implementing the changes. The Project Team manager was assigned to go out to each of the departments that would have an individual represented on the Project Team and have them identify the individual who would represent them and when he/she would be available to start working on the project. Once the availability of human resources to support the project was defined, the initial start date for the project could be set. In this case, the project set date was September 1 with the completion date 90 days later on November 29.

When the project boxed in dates could be set for major checkpoints within the project cycle, they decided to define major tollgates for management review at the end of each phase. As a result, the following five tollgates were established:

- Tollgate #3: September 12, 2016 firm date
- Tollgate #5: September 30, 2016
- Tollgate #8: October 20, 2016
- Tollgate #10: November 15, 2016
- Tollgate #12: November 29, 2016 firm date

You will note that only the first and last tollgates are considered firm dates. The Project Team will be allowed to propose alternate dates for all the other tollgates. The Project Team also will be responsible for establishing and conducting the other 10 tollgates. Targets for all 12 tollgates will be included in the project work breakdown structure.

- High-level cost estimating and budgeting
  This is usually just a very high level estimate that is best done in conjunction with the individual or team that submitted the proposed project. These estimates should be accurate to plus or minus 20 percent. A typical example for installing a new software package follows:
  - Manpower requirements
    1. Programmers: 10 employee months at $10,000 per month = $100,000
    2. Project manager: 5 employee months at $12,000 per month = $60,000

  3. Training: 6 man months at $6000 per month = $36,000 (two hours per employee in each department that will be using the software)
  4. Consultant: 5 months at $20,000 per month = $100,000
- Equipment
  1. Software: $250,000 plus $50,000 per year for five years = $500,000
  2. Computer equipment = $150,000
  3. Additional space requirements = none
- Total costs for the project = $946,000

Note: In this case, all the human resources came from two departments, Project Management and Information Technology. In cases where there are many departments represented on the Project Team, the cost estimates should be broken down by employee hours and cost per department.

- Draft executive summaries for each proposed project/program
  The Project Team manager and the project manager will prepare an executive summary for each proposed project/program. Shown in Figure 2.11 are the kinds of information that should be included in a draft executive summary for each project or program in the portfolio.

Outputs

- High-level budget set for each project
- Key management checkpoints are defined and documented
- The project team is defined and assigned
- Executive summaries are prepared for each proposed project/program

Activity #7: Obtain Executive Approval for Each Project and Program Including Its High-Level Milestones and Budget

Below are the inputs, activities, and outputs for this activity.

- Input(s):
  - Executive summaries for each proposed project/program
  - A list of proposed projects with their start and end dates and a budget for each one
  - A list of qualified portfolio leaders and project managers

Initiative: Finance – Budgeting Process	Sponsor(s): Dewey, Pleasum & Howe
Benefit Type: ☐ Quick-Win X Improvement (check one) ☐ Innovation	**Driver Type:** ☐ Revenue X Expense (check one)
Priority: ☐ Very High ☐ High X Medium (check one) ☐ Low ☐ Very Low	**Manager:** Ike Anduitt

Current-State Performance

The existing Budgeting Process takes at least 6 months, starting in May/June each year and ending with a presentation to the Board in the 4th Quarter. It is characterized as: financially driven output (business management information is developed but not published with the budget), too iterative, and provides limited downstream use due to the age of information for current year's operations. The Current State Process' costs exceed $1M.

Initiative Objective(s)

Shorten the cycle-time by at least 50%. Start the process later in the year to provide more meaningful information for business management.

Key Performance Measure(s)

Cycle-time

Change Management Considerations

There is strong user and management support to accomplish the objective; however, the internal customer at the end of the process may affect the change recommendations. Change for this process must be driven by cross-functional teams for results to happen.

Investment	Reward

People:
  Assemble an cross-functional team:
    15 people @ 3 days each
    1 to 2 full-time analysts
Technology: N/A:
Capital: N/A:
	• Cycle-time reduction; reduce process costs to a reasonable level. Assuming a 50% reduction of process costs, projected annual process savings are $500K
	• Creation of a better business management tool.
	• Better linkage to AFC process and sales and marketing execution.

Total Investment: 60 days	**Total Reward: $500K**

High Level Action Plan:	Finalize an approach using a Process Improvement Team Build on existing format Confirm with Corporate Management Communicate to Midwest Region Monitor deliverables/results
Time Frame (elapsed time): Staffing (FTE, skill sets):	4 to 6 weeks Cross-Functional team needs representation from all areas.

FIGURE 2.11

Organizational project portfolio: Draft executive summary of the project.

- Activities:
  - Group the projects/programs into subportfolios
  - Prepare a layout timeline for all projects and programs indicating their interrelationships
  - Present portfolio layouts to the executive committee and get their approval
  - Update documentation related to the project team's activities, responsibilities and authorities
- Output(s):
  - List of approved projects/programs with their final executive summaries, start and end dates, and budget for each one
  - An approved list of portfolios with project manager, project team manager, and sustaining sponsors for each project/program
  - A formal project report that summarizes their activities and assumptions
  - Minutes of the executive committee review and decisions
  - Updated set of operating procedures related to the process of selecting the projects/programs that will be included in the organizational portfolio
  - Project team manager assigned to each approved project/program
  - Start and end dates and a budget for each project
  - Key checkpoints for each project or program if applicable
  - Executive summary for each proposed project/program

Inputs

- Executive summaries for each proposed project/program
  These executive summaries will be used to help group the individual projects/programs into subportfolios. They also will be used as a key input into the final executive summary.
- List of proposed projects/programs with their start/end dates and a budget for each
  This list is primarily used to establish proper budget controls and to provide input into the final executive summary.
- List of qualified portfolio leaders and project managers
  This list includes the capability and experience related to each of the Portfolio leaders and it is a key input into how the projects/programs are combined into subportfolios.

Activities

- Group the projects/programs into subportfolios
  Now the Portfolio Development Team will review the subportfolios for the already active projects/programs. Careful attention will be paid to points in the already active projects/programs where a high activity workload occurs and how the workload drops off with time. The present subportfolios should already be grouped in a logical combination of projects/programs. As previously stated, these are normally grouped based upon the type of project/program that is being implemented. Based on the workload of the individual, additional projects/programs will be assigned to the Portfolio leader. When the number or complexity of the proposed active projects/programs exceeds what the present Portfolio leader can effectively manage, an additional subportfolio is created and a new Portfolio leader is assigned. The end result from this activity is a list of subportfolios, which includes the projects/programs included in each subportfolio and the name of the individual Portfolio leader.

  Note: On occasion, a Portfolio leader also will be assigned to become a Portfolio Development leader for a project or program. Prepare a layout timeline for all projects and programs indicating their interrelationships.

  Using the executive summary for each new project/program that includes the overall schedule of activities in the work breakdown structures for the currently active projects and programs, the Portfolio Development Team should construct a view of the total group of projects/programs that are scheduled to be active within the organization. This will allow them to determine how they relate to each other and to define overall workload and change management problems. This is accomplished by establishing a high-level Gantt chart that shows all of the approved projects/programs. Particular attention is given by the Portfolio Development Team to define inputs into the individual projects/programs that come from other active project/program. These inners timings are extremely critical and often result in reassigning priorities and/or changing schedules to accommodate these inputs (Figure 2.12).

Combined 3–Year Improvement Plan

Activity #	Activity		2017								2003					2019				Person Responsible
			A	M	J	J	A	S	O	N	D	J	F	M	2	3	4	1	2	3	4	
P	3-Year 80-Day Plan 419																					H.I.-EIT
02	Develop Individual Divisions																					EIT
										Cycle 1					Cycle 3							
BP	Business Process																Cycle 4					EIT/Bob C.
10	EPI											Cycle 2										EIT/Tom A.
ML	Management Support Leadership																					
10	Team Training																					EIT/Task Team
20	DIT																					Dept. Mgrs.
61	MEWA																					Division President
62	Employee Opinion Survey																					H.I.
30	Strategic Direction																					Sam K.
40	Performance Planning and Appraisal																					Jom B.
60	Suggestion System																					Task Team
SP	Supplier Partnerships																					
10	Partnership																					H.I. – F.M.
20	Supplier Standards																					H.I. – J.A.
30	Skill Upgrade																					Bob C.
40	Cost vs. Price																					Jack J.
60	Proprietary Specifications																					Division President

▓▓▓ = Action

▓▓▓ = Ongoing Activity

FIGURE 2.12
Sample high-level portfolio management plan.

- Present portfolio layouts to the executive committee and get their approval
  The Portfolio Development Team will make a major presentation to executive management explaining how they conducted their evaluation and providing them with an understanding of why the Portfolio Development Team selected the projects/programs to be incorporated in the Organizational Portfolio. They also need to be ready to explain why any project that was not included in the Organizational Portfolio was not selected. A typical agenda for these meetings with the executive management team follows:
    - Explain the methodology that was used to prioritize the projects/programs.
    - Review the prioritized list. Explain in detail why the first 10 projects/programs were positioned in that particular order. Invite the executive team members to question any individual program/project that was not rated in the top 10.

- Review the resource limitations that were placed upon the Portfolio Development Team.
- Review the projects/programs that the Portfolio Development Team recommends be included in the Organizational Portfolio.
- Review the three to five next lower priority projects/programs that were not recommended for inclusion in the Organizational Portfolio, explaining what additional resources would be required to have them included in the active projects/programs.
- Review the present active programs/projects that the Portfolio Development Team recommends be terminated in order to free up resources for project/program that will have a more positive impact upon the organization.
- Recommend the group of programs/projects that should be included in individual subportfolios.
- Facilitate discussion among the executive team so that they finalize a list of approved active programs/projects. We recommend that the final list is documented on a whiteboard or a flipchart during the meeting so that there is no misunderstanding later on.
- The Portfolio Development Team also should discuss the difficulties they had in completing the project and the lessons learned.

As a rule of thumb, schedule 10 to 15 minutes per approved project. Our experience indicates that these meetings do not run smoothly unless there are resources enough to cover all of the proposed projects/programs. Often even a second or third meeting needs to be scheduled before the executive team can come to a common agreement on which projects will be included during the coming years. In too many organizations, individual executives' primary interest is on "what's good for their function" rather than "what's good for the organization as a whole."

Be sure to carefully document these meetings so there is no misunderstanding of what was agreed to. This is a very important meeting and we strongly recommend that you schedule it a minimum of five days in advance so that the executive team has an opportunity to rearrange their schedule so they can attend the meeting. All too often when one of the executive team members assign someone to set in for them they disagree and take exception with the final results.

The Portfolio Development Team is responsible as well for preparing a formal document that covers, in greater detail, each of the items discussed with the executive team. This report also should include the assumptions of what the Portfolio Development recommendations were based upon.

- Update documentation related to the Portfolio Development Team's activities, responsibilities, and authorities
  If this is the first time that a Portfolio Development Team has been assigned to complete a similar project, then there are a number of documents that they should generate to assist the team that goes through the process the next time period. Typical procedures would include:
  - How to prioritize projects/programs.
  - How to determine the amount of discretionary resources that are available.
  - Roles and responsibility of the Portfolio Development Team and the Portfolio Development leader.
  - How to assess the accuracy of the projected costs and savings for the proposed projects/programs.
  - How to document the results of their activities.

If the procedures are already available, then the Portfolio Development Team should review and update them as appropriate.

Outputs

- List of approved projects/programs with their final executive summaries, start and end dates, and budget for each one
- An approved list of portfolios with project manager, project team manager, and sustaining sponsors for each project/program
- A formal project report that summarizes their activities and assumptions
- Minutes of the executive committee review and decisions
- Updated set of operating procedures related to the process of selecting the projects/programs that will be included in the organizational portfolio
- Project team manager assigned to each approved project/program
- Start and end dates and a budget for each project
- Key checkpoints for each project or program if applicable
- Executive summary for each proposed project/program

SUMMARY

In this chapter, we have focused on *Phase I: Develop the Organizational Portfolio* and how this phase frequently occurs as part of the annual budget cycle for the organization. To help the reader understand the complexities of defining and deciding which business cases are included in the Organizational Portfolio, we presented an organizational structure for a "typical" manufacturing organization. However, it could be applied equally to an organization driven by a set of business transactions or one driven by the provision of services.

In this chapter, we have provided answers to the CEO's typical questions through focusing on the development of the Organizational Portfolio, which will establish the foundation for these targeted work initiatives.

Finally, we have introduced each of the seven activities that make up Phase I. We have described how these activities (or processes) evaluate the set of proposed business cases (each one aligned with a value proposition) to determine which of them will be selected and prioritized as a project or program to be approved and included in the Organizational Portfolio. Like any process, each of these activities has at least one input, one activity, and one output that should be applied in a generally sequential fashion beginning with Activity #1 and ending with #7, either when developing an Organizational Portfolio for the first time or, if one has already been established, for the next iterative developmental time period.

Also during this phase, each of the active projects/programs had an individual assigned as the sustaining sponsor of the project/program. In addition, a project manager was named for each approved project and program. Also the programs that were scheduled to be active were grouped into subportfolios and a Portfolio leader was assigned to each.

An organization without a properly developed portfolio for its projects and programs is a lot like Buridan's ass that was equally hungry and thirsty. It was standing halfway between a pile of hay and a bucket of water and it kept looking left and right, back and forth, trying to decide between the hay and the water. Unable to decide, it eventually fell over and died of both hunger and thirst.

Bill Ruggles, based on *Buridan's Principle*
with thanks to French philosopher Jean Buridan
and American physicist Leslie Lamport.[3]

Too much effort goes into doing projects and not enough effort in selecting the correct ones.

H. James Harrington

REFERENCES

1. Meredith, J. R. 2012. *Project management: A managerial approach*, 8th ed. Hoboken, NJ: John Wiley & Sons, 47–62.
2. Ibid, pp. 65–73.
3. Lamport, L. 1984. *Buridan's principle*. (Revised February 2012) Maynard, MA: Digital Equipment Corporation, Systems Research Center.

3

Phase II: Create the OPM System Implementation Plan

INTRODUCTION

The purpose of this phase is to create an Organizational Portfolio Management (OPM) System Implementation Plan and set the stage for implementation. Also known as Project Management Information Systems (PMIS), the Organizational Portfolio Management System is a software program/application which provides a centralized system and methodology for the management of portfolios, programs, projects, resources and costs, providing visibility into overall and detailed level performance tracking. As an input to this chapter, we have a number of portfolios, anywhere from two to five (10 depending on the size, scope, and span of the organization), scalable as the organization grows (or contracts). Each portfolio has a Portfolio leader established. Each portfolio contains a number of subportfolios, projects, and programs. Each project/program has a Portfolio leader assigned to it, and each Portfolio leader is managing multiple personnel/resources, who likely are participating on other projects as well.

The problem most organizations face is how to oversee not only the Portfolio leaders, but the way the collection of portfolios, programs, and projects collectively interact and operate using shared and constrained resources. One solution is to bring them all together into a Project Office that has all of the Portfolio leaders and project managers reporting to it. Another is to have them all remain independent and reporting to the function that they are concerned with (e.g., all of the Portfolio leaders who are assigned to software programs could be reporting up through the information technology function). All of the new product development Portfolio leaders and project managers could be reporting into different departments in the development organization. Each type of organizational structure has distinct advantages and disadvantages.

Then there are the questions of portfolio operations and governance. Should all of the portfolios within the organization be operating to the same guidelines? For example, should the portfolios that are related to new products operate under the same guidelines as the portfolios that support information technology functions, or would that put too great a constraint on innovation? Each organization must come to consensus on how to agree on common operating procedures to answer, up front, these key questions. How will the outputs from projects that feed into inputs to other projects be connected? Where are the handoffs and points of negotiation? This may be fairly simple when working within a single portfolio of programs and projects, but careful consideration must be given on how to structure and manage outputs (and resource management) across multiple portfolios.

There are advantages to having a Project Management Office established with roles, authorities, and responsibilities clear to all involved. In our experience, organizations trying to manage portfolios of programs and projects without a Project Management Office typically run into conflicts early on during initiation and planning around lack of standard definitions, governing practices, and project management authority. This becomes even more critical during the stages of execution, monitoring, and control.

OPM Systems play a key role in managing the Organizational Portfolio by minimizing the potential for projects continuing to operate past the point when they should have been shut down. Small to mid-sized organizations with four or less portfolios probably do not need a dedicated Project Management Office, and the same principles can be applied by forming a committee that meets and works together to define and agree on the operating rules, reporting, and project status/review process. Portfolio management software also provides visibility for the Portfolio leader working with the program and project managers reporting to their Portfolio leader. Managing across portfolios helps the Portfolio leader add positive value in support of the organization's strategic objectives, while staying within budget and minimizing resource burdens to the operations, functional work units, and departments they are intended to support.

OPM involves identifying and aligning the organization's priorities, establishing governance, and providing a framework for performance management and continuous improvement. However, in order to successfully implement a sustainable OPM System, the approach to OPM must go beyond the aggregation of projects, programs, and subportfolios that support the organization's strategic objectives.

An organization seeking to embrace an OPM methodology must develop the framework and infrastructure for deployment of an OPM System to manage the inherent complexity that arises when the needs of the organization require strategic alignment on its collection of projects and programs. For example, a gap analysis performed recently with a large North American retailer revealed that project management previously had been largely decentralized and that the organization's strategic plan was not in alignment with the multitude of requests for projects that the senior leadership team was tasked with prioritizing for the upcoming year. A Portfolio Operations Deployment (POD) team was established that consisted of key stakeholders who were facilitated by outside consultants to assemble a complete picture of all current program and project requests. Only with the visibility of having all of the programs and projects on the board did it become clear to the senior leadership team that there were strategic decisions planned for the second and third quarter of the year that would have major implications on projects planned for the first quarter (Q1). These projects would have had major capital investments incurred in Q1, only to be suspended (and likely cancelled) in Q2 or Q3. This simple 360-degree view of the organization's programs and projects helped save the client tens of thousands of dollars (and countless resource hours) by stopping sure-to-fail projects before they started.

The remaining programs and projects were prioritized against the organization's strategic objectives using empirical criteria in a matrix similar to the example in Table 3.1.

- Rating: Programs and projects are rated 1 to 5 based on their impact on the specified criterion (established by the steering team) with 1 as low/5 as high.
- Weighting: (Optionally) each criterion can be assigned a weighted value with 1 as low/5 as high if it is desirable to give more "weight" to one or more criteria over the others (e.g., if customer growth and retention is the organization's top strategic priority, this could be weighted as a 5 and the other criteria given a 4.
- Finally, calculate the total scores for each program and project, then prioritize them to quickly demonstrate which are the top candidates for scheduling and implementation.
    * (Rating × Weight) = Criterion Score
    + Criterion Score + Criterion Score + Criterion Score = **TOTAL**

TABLE 3.1

Sample Program and Project Prioritization Matrix

Portfolio Management: Program and Project Prioritization Matrix					
Criterion	Scoring	ERP Program	Cost Containment	Profit Improvement	6 Sigma Analytics
Strategic alignment	Rating	3	5	4	3
	Weight	4	4	4	4
	Score	12	20	16	12
Customer growth	Rating	3	2	5	5
	Weight	5	5	5	5
	Score	15	10	25	25
Cost (list actual	Rating	2	5	5	4
costs as available)	Weight	5	5	5	5
	Score	10	25	25	20
Revenue (planned	Rating	2	3	5	3
or projected,	Weight	5	5	5	5
based on prior results if available)	Score	10	15	25	15
Risk, constraint	Rating	2	4	5	4
(or other	Weight	4	4	4	4
impacting factors)	Score	8	16	20	16
	TOTAL	55	86	111	88
Criterion		ERP Program	Cost Containment	Profit Improvement	6 Sigma Analytics

Our client in this example also realized the need for a comprehensive centralized OPM System to manage the complexity and need for constrained resources across the spectrum of prioritized programs and projects.

The OPM System solved the organization's need to have a centralized collection of independent projects or programs grouped together in a centralized database to facilitate its prioritization, effective management, and resource optimization in order to meet strategic organizational objectives.

ACTIVITIES FOR PHASE II

Phase II (*Create the OPM System Implementation Plan*) is made up of the following four activities:

Activity #1: Prerequisites for OPM System Implementation Planning
Activity #2: Establish the Project Management Office (PMO)

Activity #3: Assemble the OPM System Implementation Team
Activity #4: Create the OPM System Implementation Plan

Activity #1: Prerequisites for OPM System Implementation Planning

Creation of the OPM System Implementation Plan is dependent on the following prerequisite activities. Assuming the organization is just starting out with a concept of implementing an OPM System, this activity provides the required guidance.

One of the very first things that should be done is to establish an integrated Project Management Plan for each project that includes the following sub-plan elements (Figure 3.1) for effective portfolios.

1. Project Integration Management: Answers the question: Where does this portfolio intersect with its subprograms and projects, or with other portfolios? (Critical to understand and minimize cross-portfolio impacts.)
2. Project Scope Management: The span, scope and boundaries of each project or of the portfolio program (as outlined in the Business Case). Ideally the Project Scope also includes any items that have been identified as out of scope.
3. Project Time Management: Outlined in the Work Breakdown Structure; see next section.
4. Project Cost Management: As outlined in the Value Proposition and Business Case. Derived from the high-level budget outlined in the Value proposition and Business Case, integrates Earned Value Management (EVM).
5. Project Quality Management: Essential to establish how project performance will be measured and evaluated.
6. Project Human Resource Management: Consider use of the Critical Resource Chain method (A Framework of Critical Resource Chain in Project Scheduling[1]) to maximize and integrate constrained resources within a tight schedule.
7. Project Communications Management: See sample plan in the next section and Chapter 5 (Practical Applications of Change Management within Portfolio Management).
8. Project Risk Management: See sample plan in the next section; also consider the processes outlined in the PMI's Practice

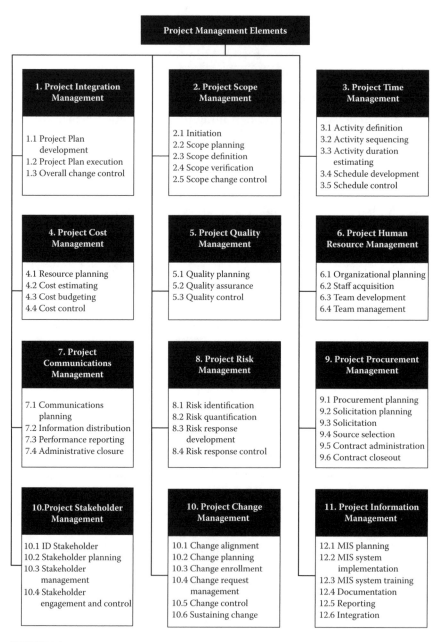

FIGURE 3.1

The 12 Project Management Elements for Effective Portfolio Management.

Standard for Project Risk Management,[2] which outlines risk identification, risk quantification, risk response development, and risk response control.

9. Project Procurement Management: This may be handled by a dedicated procurement department within the organization, or within the province of the Project Management Office on a dedicated or as-needed basis for individual portfolios, programs, and projects.

10. Project Stakeholder Management: Stakeholder Identification and Engagement is particularly critical across the portfolio as the Stakeholders may very well have differing and often conflicting objectives.

11. Project Change Management: As of this writing, work is under way to introduce Project Change Management within the PMI methodology.

12. Project Information Management: See Activity 4: Sample OPM System Implementation Plan: Section 4. Technology Resource Management Process; Sample Documentation. This element is covered for organizations undertaking PMIS/OPM System implementation.

Activity #2: Establish the Project Management Office (PMO)

Doing a good job at this stage is absolutely crucial.

H. James Harrington

Establishment of the PMO is critical for the success of the OPM System as an organizational unit with authority to affect change (within the boundaries of approved business cases) and serve as central repository for all project artifacts within the organization.

As directed by the Executive Committee, the organization's PMO is there to not only lead the projects, but to serve as the face of the organization's change agenda, implementing and integrating the OPM System into existing organizational business processes, including planning, auditing, technology optimization, process optimization, sales, marketing and new product, and service development. The PMO aids in the identification and training of portfolio, program, and project managers, and is an essential actor in the implementation of both the OPM System and the Program Management infrastructure the system is intended to support.

Distributed Portfolio and/or Program and Project Management scenarios in larger organizations may require one PMO for each location. However, if there are separate PMOs for different functions, communication, cooperation, and alignment must be established between them, and the Project Team manager and senior executives must give the Portfolio leader the authority to unify them and bring them together. The OPM System becomes a powerful tool for attaining this required level of synergy across multiple distributed locations. The following high-level tasks should be considered prior to embarking on wide-scale deployment of an OPM System, because the PMO provides a central authority for managing the span of portfolios, programs, and projects across the enterprise.

1. Direction for the establishment a PMO and related activities (or an interim PMO if one does not exist)
2. Establishment of OPM System implementation milestones
3. Outlining of the organizational structure (via org charts) and drafts of job descriptions for the permanent PMO, which defines the roles, responsibilities, and authority for a PMO and the Portfolio leaders
4. Drafting of PMO policies and procedures and reviewing them with the Executive Committee
5. Development of Standard Operating Procedures (SOPs) for the interim (if needed) and permanent PMO
6. Standardization of Portfolio/Project Quality Assurance and Quality Control Processes and PMO templates (see Appendix C)
7. Establishment of permanent PMO measures and reports

At this stage, the organization should be giving serious consideration to chartering a PMO (if one is not already in place). There are many advantages to having a PMO, including:

- Management has a central authority and single point of contact for real-time information, reporting and status on the priorities, and milestones of all programs and projects.
- The PMO has a total view of all the projects and will be the first to go to management to warn when a project should be canceled.
- The PMO highlights projects that are having problems and gets management to react to the problems that require their attention.
- The PMO provides a central contact point to understand what projects are being considered for approval and those not yet ready for official project status.
- The PMO can help prepare value propositions and business cases, and are ultimately held accountable for any overruns in budget, ability to

meet schedules, or lack of performance in any of the departments within its projects.

- The PMO plays a difficult role in negotiating project deliverables and tasks with a variety of operations and development managers.

In a recent example from our work with the Office of Innovation and Information Technology (OIT) leadership team and the Program Management Office (PMO) at Nova Southeastern University (NSU) (Fort Lauderdale, Florida), a Strategic Planning Team was chartered in tandem with their Lean Six Sigma Green Belt and Black Belt training to develop a Strategic Planning process that yielded their PMO mission statement aligned to their vision and values.

- **PMO Mission Statement:** The mission of Project Management Office (PMO) provides an enterprise-wide approach to identify, prioritize, and successfully execute a technology portfolio of initiatives and projects that are aligned with the NSU strategic goals and educational vision. A primary responsibility is to manage and control project constraints by ensuring project plans are implemented on schedule, within scope, and budget. Project management leadership is responsible for establishing and implementing best practices for the benefit of NSU in a way that encourages collaboration, standardization, and overall improvement in our educational community.
- **PMO Vision Statement:**
  - Build project management maturity at the organizational level.
  - Support students, faculty, staff, and the NSU community as a source of project management leadership and expertise.
  - Promote best practice standards, quality, and methodologies into a project management discipline.
  - Utilize PMBOK-based methodology[3] as well as support "best fit" approach for project management at NSU.
  - Provide a channel of communication for project status, financial health, and mitigation of issues, risk, and dependencies across projects, departments, and/or divisions.

Examples of Project Decentralization

Organizations that grow to the size where PMOs and OPM Systems become warranted and justified expenditures often are at a size, breadth, and scope that necessitates project development be geographically and functionally decentralized. With corporate and regional headquarters, branch locations, hubs, and offices (and home offices) in play, the system must adapt to a dispersed and flexible workforce, distributed regionally, if not nationally and internationally around the globe.

In a 2007 self-audit of the Texas Department of Transportation (TxDOT) Field Operations, several essential elements for decentralized project management were recommended. The first recommendation focused on the deployment of standardized tools and strategies to manage their large portfolio of programs and projects and subportfolios. According to the audit report, "Developing standardized strategies and tools for managing project schedules throughout the project development life cycle would improve execution of the planning process for future projects and improve the accuracy of reported schedule progress. Potential issues could be identified early in the development process" so the execution strategy could be adjusted to ensure milestones are achieved.

Following the release of the audit report, the administration set up a working group to identify and recommend a statewide OPM System for use during project development. They recommended the use of Critical Path Method (CPM) scheduling and implementation of Oracle's Professional Project and Portfolio Management software application.

Although CPM scheduling is not uncommon for transportation project development, its use at the portfolio level of the enterprise had been uncommon in the public sector, in part because CPM scheduling requires trained employees with skills in both (industry specific) project development and formal project management.

In addition, the TxDOT work group recommended and implemented an interim Project Management System for all projects in active development, and an interim PMO to give them the best chance of meeting the key milestones identified and achieving their strategic objectives.

Activity #3: Assemble the OPM System Implementation Team

Establishing clear roles and responsibilities is crucial at this stage.

H. James Harrington

While in most cases the Portfolio leaders do not directly manage the employees that roll up under their programs and projects, they are directly responsible for the time to which various resources are committed throughout the project and delegating the activities described in the work breakdown structures across the series of programs and projects under their span of control. The Portfolio leader also should be highly capable of motivating individuals to complete work when the individual may not always see the big picture or rationale for the importance of the given tasks and roles. The Portfolio leader can explain that the individual's completed work may be required for other team members to complete their tasks. The Portfolio leader reviews and provides input on program and project plans prepared by other project managers who work within the portfolio of their responsibility, such as training and mentoring programs and project managers to improve their knowledge and base of experience, thereby increasing organizational capacity.

There is a certain subtle art in mentoring project managers working on your portfolio without giving the impression that you are dictating to them or circumventing the authority of their existing management structure that provides the structure for their daily work (their perceived "real job").

In order to ensure the success of the organization's key objectives, the organization's professional project managers should have a defined career path toward becoming a Portfolio manager. In order to ensure the success of the organization's portfolios, programs, and projects, the Portfolio Steering Committee and all aspiring Portfolio leaders also must be able to answer—or find the answers—to the following five essential questions:

1. How does our portfolio align with the organization's strategic objectives? (This should be clear at this stage.)
2. Who are the primary stakeholders of our portfolio and how will the results and status of the portfolio be communicated and reported? (This should be included in the stakeholder register.)
3. What levels of training, experience, abilities, and knowledge should be required of potential Portfolio managers?
4. What is the current baseline skill level of the current project managers? Is the performance delta between project and Portfolio leaders known and understood?
5. How do you evaluate individual project managers and give feedback to them when you are not their manager? (Consideration should be given to both the interpersonal and technical results of their project management activities.)

Corporate culture is shifting to support collaboration in a less isolated work environment. They see the value and advances in productivity they get from making their workers more comfortable with each other, building trust, and, ultimately, making the company a lot more money in its respective market.

For additional information on the roles/definitions of portfolio, program, and project managers and associated teams, see Appendix A: Project and Portfolio Management Definitions; Organizational Influencers on the Project Lifecycle. At a high level, Portfolio Managers manage portfolio components, programs, and projects against strategic objectives. Program managers manage programs and projects that are best managed when grouped collectively, and project managers plan and manage temporary endeavors to achieve specific requirements versus the constraints of time, cost and resources.

Activity #4: Create the OPM System Implementation Plan

The section that follows is a detailed sample OPM System Implementation Plan that can be used as a template for outlining milestones and project artifacts. A sample organizational chart is provided, as well as an outline of scope and responsibility, and Project Management and Change Management processes (including scheduling, budgeting, assumptions, dependencies and constraints, risk management, change management, and control).

Then, the Technology Resource Management process outlines a sample system environment, methodology, and support systems overview and common documents generated during the OPM System Implementation Project.

There are certain decisions that will impact the way the organization adopts, utilizes, and supports the OPM System. For example, the decision is to go with a traditional client server-based system maintained on internal servers in-house, or utilization of Software-as-a-Service (SaaS)-based products that provide greater degrees of flexibility and scale, typically at a lower cost.

We see adoption of SaaS from on-premise as a way to gain faster access to key functional IT PPM capabilities without maintenance or internal labor costs and with decapitalization benefits.

Application Life-Cycle Management Market Data Analysis; International Data Corporation (IDC), December 2013

Finally, the sample plan concludes with resource and work management, and budget and resource allocation, including a sample cost/benefit analysis.

Sample OPM System Implementation Plan

1. Introduction
   1.1 Executive Overview
   The purpose of the OPM System Implementation Project is to deploy OPM System software, policies, and procedures, including alternate/optional components including organizational structure and reporting. The software will be tested for three months on a sample program to ensure all applicable features are enabled and provide lessons learned, which will be used as the basis for portfolio/program/project manager (PM) training on the new system. The software has an existing per license startup cost, plus additional costs to be determined for vendor configuration and consulting support.
   The objectives of this initiative are to:
   A. Procure and implement a OPM System
   B. Implement project time tracking and resource management
   C. Create standard reports within the overall PMO governance structure
   D. Elicit additional requirements for eventual integration with other back-office systems (including but not limited to Active Directory, HR, and Finance)
   1.2 Milestones
   1. Create the database and central archive/project repository
   2. Gather input from the test (process execution and end user input)
   3. PM training (process execution and usability review)
   4. Time tracking workshops
   5. Completion of Change Management phase (including enrollment, communication, training, documentation, and end-user support)
   6. Reporting (including milestone reporting, portfolio roll-up, resource, time, and task detail)
   7. Implementation and Integration

1.3 Glossary

Optionally, this section can provide standard operational definitions for common technical terms and acronyms incorporated in this plan.

2. Project Components

2.1 Project Artifacts

Use this section to manage the artifacts intended to be produced and/or utilized throughout the scope of the project, inclusive of the following components (Table 3.2).

2.2 Portfolio Management Organization

Sample organizational chart (Figure 3.2).

2.3 Scope and Responsibility

The sponsor, in alignment with the organization's strategic objectives, has allocated budgeting and management authority for this initiative. This project is within the span of control of the organization's Program Management Office (PMO), and a dedicated project manager (PM) has been appointed. Subject Matter Expertise (SME) and end-user testing (UAT) will be utilized to failsafe the implementation of the OPM System. System manuals will be provided by the vendor; end-user training, documentation, and job aids will be developed by the Change Management/ Training Team in coordination with Information Technology and Systems Support.

TABLE 3.2

Project Artifacts

Component	Start Date	Completion Date (Planned)	Responsibility
Project plan	Q1	Q1	Project Team Manager, Portfolio/Project Manager/PM
Action log	Q1	Q1	PM
Design specification	Q1	Q1	Business Analyst (BA)/PM
Software installation amd database administration	Q2	Q2	Systems Administrator, DBA
Initial pilot report	Q2	Q2	Project Team
How to guide	Q2	Q2	Change Management Team
Implementation plan (production)	Q3	Q3	Project Team Manager, PM
Program review of lessons learned	Q3	Q3	PMO Manager, Project Team Manager, PM

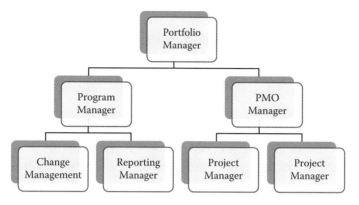

FIGURE 3.2
Sample Portfolio Management org chart.

2.4 Implementation Responsibility

Utilize this section to expand upon the sample roles and outline responsibilities as appropriate throughout implementation's lifecycle:

- Project Team Manager: Secures implementation-required resources and provides technical and political direction to individuals working on a specific project/program
- Project Manager: For implementation of the OPM System
- Change Management, Communications, and Training Team
- Deployment Team Members: With resources and approval to carry out required tasks
- Database and System Administration: For support and maintenance of the system
- Information Technology and Systems Support: For general ongoing user support

(See Appendix A: Project and Portfolio Management Definitions; Organizational Influencers on the Project Life Cycle)

3. Project Management and Change Management Processes

3.1 Schedule

The schedule for this project is designed for concurrency and speed of implementation (in under a year). Driving factors include planned expansion into new markets that will drive exponential growth and new project requests. Project Teams will meet biweekly at first (Monday a.m. goal setting, and Friday p.m. weekly status reviews). A third Wednesday mid-week status meeting will be inserted at the first sign of project slippage. Sponsor debriefs will be held every other week or as needed on an ad hoc basis. Project status reports will be uploaded to the central repository each week, and made available as needed for executive and Portfolio Steering Committee monthly and quarterly reviews.

3.2 Budget Review

The project manager is responsible for adherence to the project schedule and budget. Variation +/– 5 percent of the project budget and/or schedule requires immediate review with the Project Team manager.

3.3 Assumptions

The only assumptions are that the project will continue to be funded (and staffed) as long as it stays on track within the published schedule. Barring major financial or environmental impacts, the project is intended to be completed on time, within budget.

3.4 Dependencies and Constraints

The implementation of the OPM System is dependent on alignment of the organization's strategic plan 2014–2015. The only other dependency is the decision on optional system integrations (with Active Directory, HR, and/or Finance).

The existing constraints include the organization's capacity for planning and prioritization of projects to be managed through the OPM System. A minimum of one program or project is required for the pilot phase. Once the project is successfully under way, time tracking will be enabled for the end-user testing phase.

3.5 Risk Management

The Project Team manager is responsible for ensuring the Project Team identifies, prioritizes, and mitigates any known or possible risks to successful completion, as well as the ongoing sustainability of the system. The known risks to date are as follows:

1. Senior management commitment: This project cannot proceed until this risk is mitigated; it will require alignment within the ranks of the Executive Committee, before any vertical alignment is attempted with the rest of the organization.

2. Alignment with Strategic Objectives: Implementation of the OPM System will require changes to the way the organization

adapts to and aligns its objectives to support the overall strategic plan 2014–2015.

3. Resistance to change: Project managers will need to actively involve their teams, subject-matter experts (SMEs), and change agents (to be identified) to ensure their enrollment with the change effort to minimize resistance to change. This system and its outcome reports will only be as good as the quality of input required to get actionable information up to senior management.

4. Technical and environmental constraints: Contingency plans must be in place for system/database redundancy off-site at a secured location. Within the Wide Area Network (WAN) and Local Area Networks (LANs), access to the system is to be always on to minimize impacts to the project schedule. Service levels will be established to ensure 99.9 percent system uptime within normal business hours including Saturdays, with an agreed upon (tentative) maintenance window of Saturday from midnight until 2 a.m. for planned/scheduled downtime for database backups and system maintenance.

3.6 Change Management and Control

Change Control Board: In Information Technology and software development, a Change Control Board (CCB) or Change Control Committee is a committee that makes decisions regarding whether or not proposed changes to a project should be implemented, evaluating potential impacts to stakeholder groups or other systems. In many cases, the CCB also reviews results of implemented changes, and evaluates corrective actions in the event planned changes have any adverse impacts.

The project manager reports to the Change Control Committee on a weekly basis to inform about any planned actions that may impact the assembled stakeholders from the various technical and functional units/teams. The Change Control Committee will come to consensus on planned changes impacted by or impacts to the implementation of the OPM System. The project manager is responsible for updating the project plan and risk register as impacts are identified and reporting back to the Project Team manager if there will be any impacts that could impact the project schedule. The project manager will report back to the Change

Control Committee on the status of all scheduled updates for the purposes of corrective and preventive actions that will ultimately guide the implementation of the system across the enterprise.

The OPM implementation team (in coordination with the change management team) accordingly reports back to the Change Control Committee on all status relating to the implementation, end user acceptance, and adoption of the final implemented product.

If impacts are above tolerable thresholds at any point in time during the implementation, the project manager and the Change Control Committee will enact emergency rollback and recovery to ensure no other systems or functional user groups are impacted by changes determined to be problematic. The Project Team manager is ultimately responsible for the successful implementation of the system across the enterprise of portfolios, programs, and projects.

4. Technology Resource Management Process

4.1 System Environments, Methodology, and Support Systems

This section can be utilized to outline the technology environments, platforms, and support methodology of the organization.

- Computer/Network environment: For example, Windows, Cisco LAN, SQL Server
- Planned interface with other systems: For example, Active Directory, HR, Financials
- Project repository and storage policy: For example, SharePoint client-server or cloud-based
- Implementation and Support Methodology: For example, Agile, ITIL, ISO 21500

4.2 Sample Documentation

The following are just a sampling of the type of documents/ project artifacts that may be outcomes of the OPM System Implementation Plan:

- Strategic Plan (as reference and guidance)
- Project Plan (including implementation, risk and change management, communication, training, documentation, measurement and control plans)
- Project Status Reports and Milestone Reports
- Process maps (before and after), models, and design documentation
- Change request templates and procedures

- Project Team charter, scope (required), and mission statement (optional)
- Project budget, costs/benefit analysis, resource and schedule estimate
- System specifications and system manuals (vendor provided)
- User guides and system architecture (including systems integration plans if applicable)
- Release documentation
- Defect and corrective action reports
- Project lessons learned and closure report

4.3 Technical Support Resources

Use this section to outline any additional technical and/or support requirements, resources, schedule, and budget for project technical resources.

5. Resource Management, Deliverables, and Budget Estimation

5.1 Work Breakdown Structure (WBS) provides a graphical depiction of activity sequences, decomposed to an appropriate level for facilitating estimation

This section describes the estimated schedule for the near term. Best-case/worse-case range of estimates and prior project results can provide essential inputs into the initial project schedule for the project tasks and activities. Upon construction of the WBS by the team to further aid in building the Gantt chart, as well as estimating activity cost and duration planning for task concurrency and project milestones for management, Quality Assurance, tracking, and reporting purposes.

5.2 Resource Enrollment and Alignment

This section is used to outline the resource matrix, establish the Executive Committee, create enrollment session agendas, and schedule individual enrollment plans. In addition, the resource matrix can be used to plan for the anticipated skills, estimate the number and type of resources, while providing further clarity on task duration estimates and training needs of the project teams and impacted resources (see the Resource Enrollment/Alignment Matrix below, Figure 3.3).

5.3 Budget and Resource Allocation

This section describes the project resource costs, project budget, and also can allocate portions of the budget to distinct project components, such as hardware, software, training, and other

Task Name	Resource Names	Deliverable Status
Enrollment Plan		
Create Resource Matrix		
Identify Change Sponsors	Executive / PMO	Short-list of sponsors / potential Steering Committee members
Identify Change Agents and Stakeholders	VP / Sponsor	Listing of impacted Stakeholders and potential Project Managers
Identify Additional Impacted End Users	Unit Manager	Stakeholder Management Plan DRAFT
Identify Pockets of Resistance	Unit Manager	Risk mitigations added to Risk Register
Charter Steering Committee		
Create Charter (Project Summary, Scope, Goals)	VP / Sponsor	Project Charter DRAFT
Schedule Change Sponsor Training	PMO	Schedule for sponsor training and communication memo
Create Enrollment Session Agendas		
Outline Purpose and Benefits	VP / Sponsor	Purpose and Benefits Aligned with Strategic Objectives
Describe Challenges and Consequences	VP / Sponsor	Agenda DRAFT for Enrollment Session(s)
Delegate Roles and Responsibilities	VP / Sponsor	Resources for Project Plan tentatively identified
Stakeholder Testimonials	Change Agents / PMs	Approved draft of stakeholder testimonials
Announce Change Agent / Manager Training	VP / Sponsor	Communications memo including training outline and purpose
Schedule Individual Enrollment Sessions		
Outline Purpose and Benefits	VP / Sponsor	Review Purpose and Benefits Aligned with Strategic Objectives
Describe Challenges and Consequences	Change Agents / PMs	Update documentation of existing risks and potential mitigations
Explain Change Sponsor Responsibilities	Change Agents / PMs	Org chart of project reporting structure
Explain Individual Roles	Change Agents / PMs	Consensus and buy-in on roles and responsiblities
Managing Objections	Change Agents / PMs	Updated issues list / Barriers & Aids Matrix
Share Training Plan and Next Steps	Change Agents / PMs	Validated Training Plan
Obtain Stakeholder Input	Change Agents / PMs	Validated Enrollment Plan and Project Deliverables
Incorporate Input Into the Project Plan)	Change Agents / PMs	Updated Master Project Plan

FIGURE 3.3
Resource Enrollment/Alignment Matrix.

Cost-Benefit Analysis

BENEFITS (annualized)

PMO Process efficiency savings	$50,000.00
Failed Project loss prevention	$200,000.00
TOTAL BENEFITS	**$250,000**

COSTS

Portfolio Management System Software (option 1)	$10,000.00
Internal resource time and labor costs	$5,000.00
Consultant fees	$10,000.00
Portfolio Management System Software (option 2)	$50,000.00

TOTAL COSTS (option 1)	$25,000.00
3-year Net Present Value (NPV) assuming .05 interest	**$79,365.00**
Benefit-Cost Ratio	9 to1

FIGURE 3.4
Cost-benefit analysis.

direct costs. Consider inclusion of a cost/benefit ratio as well as an estimated return on investment (ROI) (Figure 3.4).

*Net present value (PV) represented as

$$PV = \frac{\text{Future Value}}{(1 + \text{interest rate})n}$$

TABLE 3.3

Project Plan Revision History

Version	Description	Date	Updated by:
1.0	Portfolio Management System Implementation Plan DRAFT	6/15/14	C. Voehl
1.5	Initial Portfolio Steering Committee review and feedback	9/16/14	J. Harrington
2.0	Plan finalized and submitted to Change Control Board	11/22/14	C. Voehl

where n = # of years to achieve the baseline (generally 3) or

$$FV = PV(1+R)n$$

6. Supplemental Section

Depending on the size and complexity of the project, this section may be used to include any number of supplemental plans and/or links to reference materials used during the execution of the project, such as alternatives analysis, vendor management plans, rollback plans, data backup and storage information, competitive solicitation criteria, install docs, failover plans, and ongoing application maintenance, security, and support requirements.

7. Appendices

As indicated in the preceding Supplemental Section, detailed documentation (numerous pages of text) on large-scale deployments, such as requirement specifications and extensive plans, may be best suited for inclusion in a separate appendix at the end of this document.

7.1 Appendix A

7.2 Appendix B: Project Plan Revision History (Table 3.3)

REFERENCES

1. Liu, S. S., and K. C. Shih. 2009. A framework of critical resource chain in project scheduling analysis. *Construction Management and Economics* 27: 857–869. Online at: http://www.iaarc.org/publications/fulltext/A_Framework_of_Critical_Resource_Chain_in_Project_Scheduling.pdf
2. Project Management Institute. 2009. *Practice standard for project risk management.* Philadelphia, PA: Project Management Institute.
3. Project Management Institute. 2013. *A guide to the project management body of knowledge (PMBOK guide)*, 5th ed. Philadelphia, PA: Project Management Institute.

4

Phase III: Implement the OPM System

INTRODUCTION

The Organizational Portfolio Management (OPM) System typically spans many functional areas involved in the project development and execution phases. The organization also must address technical barriers as well as cultural resistance to change that most organizations face when undertaking the type of sweeping changes OPM Systems are intended to affect.

Technical barriers to change will be addressed by the implementation of an OPM System. Many excellent applications, computer programs, and even simple spreadsheets and dashboards are available at varying price points depending on your budget (see Appendix B).

The implementation of the OPM infrastructure will require its own change agenda, including enrollment planning, communication, training, and reporting. The benefits, challenges, and consequences of implementing an OPM system will require constant vertical realignment and engagement by senior leadership.

The goal of the OPM System is to provide an integrated seamless infrastructure for project/program/portfolio management. This provides both internal and external stakeholders with real-time data on the status of all initiatives within the span of the system, so that corrective action and resource allocation can be taken on projects and programs before they fail (or fail to achieve their stated goals and objectives).

The implementation of the OPM infrastructure is followed by the estimation of costs, calculation of proposed net present value, and selection of projects. This is followed by the individual planning and scheduling of project and portfolio resources during their life cycle.

Once the OPM System becomes operational, it will contain hundreds if not thousands of projects, schedules, calendars, budgets, resources, tasks, assignments, reports, and other artifacts that serve as reliable data to provide real-time decision support capabilities and aid in future project resource and cost estimation.

Once the majority of high impact (and impacting) functional areas of the organization are represented, including their projects, programs, and portfolios archived in the system, sustaining a standardized OPM Systems approach becomes a valuable tool for realizing the organization's value propositions.

The functional OPM System should enable a long view of OPM, serving as the integration engine for the organization to develop new programs and projects based on approved business cases, and manage up to a 5- to 10-year portfolio of active projects and programs. The OPM System provides continual review capability to help the organization adapt to changing resource requirements, fiscal constraints, and (meet) revenue projections.

In addition to providing information to aid and guide the organization's senior leadership in the implementation and management of approved projects, programs, and portfolios, the OPM System provides project managers and team members with a data source to monitor, measure, and report on the progress of their projects; react to issues impacting completion of milestones; and report on the successful completion of project milestones once achieved.

OPM SYSTEM CRITICAL SUCCESS FACTORS

Unlike many functional work groups (that often have trouble seeing the forest for the trees), independent PMOs can quickly identify cultural, procedural, and technical challenges that require immediate resolution. These organizational challenges typically manifest through a lack of adherence to Project/Program/Portfolio Management standards and best practices, and are resolved via focus on the following activities.

ACTIVITIES FOR PHASE III

Phase III: *Implement the OPM System* is made up of the following ten activities:

Activity #1: Develop a clear vision of the organization's strategic goals and objectives

Activity #2: Communicate the change agenda: goals, objectives, benefits, risks, rewards, and challenges

Activity #3: Identify impacted business processes

Activity #4: Provide for planning and implementation phase information technology (IT) support

Activity #5: Develop universal and tailored training

Activity #6: Develop measurement and reporting standards

Activity #7: Identify risks and technology constraints

Activity #8: Schedule and facilitate user acceptance testing and end user training

Activity #9: Develop project/portfolio security and data integrity procedures

Activity #10: Implement the OPM System and report progress

Activity #1: Develop a Clear Vision of the Organization's Strategic Goals and Objectives

> This is the most important activity in this phase as it sets the stage for everything that follows.
>
> **H. James Harrington**

In very few programs today are all the resources available when needed, i.e., without creating conflicts with other activities going on within the organization. Most organizations are already operating lean, having progressively removed most of their excess capacity. This can create a major scheduling problem when projects are added to the normal workload. In order to avoid these potential pitfalls, organizations must develop a clear vision of their strategic goals and objectives by adhering to the following best practices:

- Obtain consensus on the specific need for (and benefits of) implementing an OPM System.
- Outline the policies and procedures for project and portfolio resource management.
- Create an environment of consistent and timely communication of strategic objectives, goals, needs, challenges, rewards, and benefits of the system across the organization.

- Create or modify policies for use of the OPM System.
- Foster a shared understanding of the OPM System implementation policies, goals, and objectives.
- Instill the discipline and structures to support project and portfolio cost management.
- Mandate adherence to the stated goals of the selected portfolio schedule management methodology, e.g., critical path method or critical resource chain method. We prefer the critical resource chain method because, in most cases, the availability of resources is a key consideration in managing most projects.

As an example of world-class vision deployment, the Yaskawa Electric Group has established its 2015 vision[1] of using technology to solve emerging global problems, such as the aging society in advanced countries, as well as global environmental and energy issues. Established in Japan in 1915, Yaskawa Electric now has bases in Europe, Asia, and the United States. The company has articulated and communicated its vision through a variety of media focused around its core principle domains (Figure 4.1).

Once the vision has been defined, it is now a matter of clearly communicating the vision to both internal and external stakeholders.

Activity #2: Communicate the Change Agenda: Goals, Objectives, Benefits, Risks, Rewards, and Challenges

The Executive Committee (or its equivalent) must, at the onset of the initiative, communicate the change agenda (the goals, objectives, benefits, risks, rewards, and challenges of the OPM System implementation) to impacted internal (and if required external) stakeholders.

The change agenda is both a figurative as well as literal agenda of events that gets distributed as part of the Enrollment Plan to all prospective change agents. It sets the stage for meetings or conference calls held to explain the purpose, scope, and milestones of the project; engage them in a dialog to enroll them in the effort; explain their roles; and show them how they will be active participants in crafting the plan for change.

At this stage, the Steering Team marshals required resources, such as Human Resources (HR), the Program Management Office (PMO), and corporate communications, to enlist assistance in completing the following activities that comprise the change agenda and its Enrollment/Alignment Plan:

FIGURE 4.1
Yaskawa Vision 2015. (From: http://www.yaskawa.co.jp/en/vision/index.html)

- Development of OPM policies (by HR and the PMO)
- Development of presentation materials and communication memos to introduce the new OPM System and policies (by Communications and the PMO)
- Scheduling of communication/training sessions (classroom-based if possible, virtual if not) to outline the objectives, benefits, risks, rewards, and challenges of the new system

- Demonstrate horizontal leadership alignment and vertical adoption
- Communicate benefits of the system including balanced resource allocation and fulfillment of the organization's strategic objectives in alignment with shared mission, vision, and values
- Describe challenges and consequences
- Discuss objections and resistance to change
- Explain Project Team Manager Program and project manager and Portfolio leader responsibilities
- Delegate remaining roles and responsibilities: outline the roles, responsibilities, and accountability structure (via org chart review)
- Develop change agent and Manager Training Plan and program for impacted stakeholders
- Joint development of the communication and enrollment plan with impacted stakeholders

Activity #3: Identify Impacted Business Processes

In tandem with communications development, the identification of all impacted functional groups and their business processes should be done concurrently. This work continues throughout the Organizational Portfolio life cycle, but starts here at the onset of the OPM System implementation.

- Attain consensus on standard process documentation templates.
- Document processes for common OPM System activities including CPM (critical path method) scheduling or PERT (program evaluation review technique) analysis.
- Ensuring portfolio, program, and project templates and resultant schedules adhere to the organization's strategic goals.
- Consider project decentralization factors when developing schedules and work breakdown structures for geographically diverse populations impacted by the change.
- Schedule planning sessions and meetings with subject matter experts and process owners in the development of standard scheduling templates, providing flexibility for local variation.
- Identify areas of common critical path concern, such as utilities, environmental factors, and federal government regulations.
- Review and validate business processes, models, and milestones.
- Consider the use of Monte Carlo simulations, where simpler models do not fully capture the varying permutations and multiple scenarios

in calculating resource and time estimates. Monte Carlo models are used to derive approximate solutions through statistical sampling. First conceived at Los Alamos National Laboratory, the method was named after the Monte Carlo casino, where a card player (researcher Stanislaw Ulam's uncle, to be exact) could utilize repeated random sampling to calculate results.

- Consider the use of expert judgment (consulting with an expert experienced with the type of project and system being implemented) in crafting the templates for use in each functional area on the critical path.
- Ensure the identified experts attend the communication/training sessions and have the required experience in the scheduling method selected (i.e., critical resource chain method or critical path method).
- Review the critical path for each approved project, ensuring the scheduling templates account for all items to prevent downstream churn and cost of reworking templates after the initiatives begin.

Activity #4: Provide for Planning and Implementation Phase Information Technology (IT) Support

No project should begin to be implemented unless it is part of the approved Organizational Portfolio. Therefore, the proposal for an OPM System requires its own business case. Once that is done and it has been approved, it should be assigned to a Portfolio leader to get it implemented. The IT support requirements are generally included in the proposal to establish an OPM System, and the business case includes any needed IT support (as best estimated at that time). This information now needs to be extracted from the business case and built into (and scheduled into) the OPM System Implementation project plan, to ensure IT resources (as well the other needed resources) are factored.

There are three phases to getting the OPM System implemented:

Phase 1: Getting the OPM System approved, funded, and manned
Phase 2: Implementing the system
Phase 3: Maintaining the system annually

Consider that it may take as long as a year to get started, i.e., getting the system approved, funded, planned, manned, and documented. The better part of the second year may be spent piloting and implementing the system, where for the first time the projects are put into portfolios. From year three onward (Phase 3), it becomes a matter of a regularly maintaining the system and the portfolios, programs, and projects. The OPM System helps prevent problems by centralizing the resources and providing the infrastructure for managing the diverse resource demands of the organization's projects, but projects still fall into trouble, causing programs to fall behind schedule, while others are completed, and resources can be reallocated from completed/cancelled projects to the areas of greatest need.

The Executive Committee must ensure adequate IT resources are available during the planning and implementation of the OPM System:

- Guarantee the review of all other initiatives during the planned implementation phase to identify potential impacts and resource constraints.
- Involve the organization's IT Change Control Board and PMO early on in the planning process to ensure other planned initiatives and impacted functional groups have advance notice of the changes and impacts resulting from implementation of the OPM System.
- Ensure that IT has the required staffing, technical expertise, and budget to support the implementation of the OPM System.
- Consider the use of consulting support from the OPM System vendor selected, to aid in the deployment planning process (as well as the creation of scheduling templates and training materials for IT, technical support staff, and end users).
- Plan for contingency staffing support as a temporary stop gap for additional IT infrastructure, Project Management and/or technology support resource needs, including limited retention into post-implementation support of the OPM System.
- Plan for integration of existing systems and processes, including the IT Change Control System and Technology Support System (e.g., remedy) into the OPM System.
- Involve the organization's technology support, PMO, and systems training functions in a deep review of the OPM System fields, functions, and forms, as well as configuration, layout, views, reports, data input requirements, and export options.
- The information gathered during these sessions may yield additional detailed technical input on the proposed data input and quality control methods, as well as the selected scheduling methodology and templates before they get published.

- The information provided during these sessions becomes the basis for PMO/Portfolio and project reports, technical support scripts, job aids, and end-user training materials.
- Part of the agenda needs to be dedicated to finding out the answers to some key questions including:
  - What type of reports will be required?
  - How will progress be tracked?
  - Who will be running the reports and doing the tracking?
  - What type of information (potentially via dashboards and/or reports) goes to the project team, project manager, the Project Team manager, the senior management, and sponsor levels of the organization?
  - How can the OPM System facilitate a big picture view of the span of initiatives across the enterprise?

Typically, an executive dashboard is developed so the Executive Team members don't have to receive a number of individual reports. They need to have a summary report that focuses them on the projects that need attention immediately. For example, an executive dashboard would show all the projects with a separate line item for costs, a separate line for schedule (and, in some organizations, a separate line for output performance across the top). The dashboard should facilitate either monthly or weekly status review, using easy to interpret RGY (red, green, yellow) visual controls. Red would indicate a problem, yellow indicates caution of slipping schedule or going over budget, green indicates everything is on schedule.

Areas in yellow/caution or red/warning status require the Portfolio leader to know the status and be prepared to brief the executive team on delays or problems and proposed solutions. At a minimum, the Executive Team should be briefed once a week to review the status of all the projects and the action plan to get any of the projects that were in red back into the green. Intraweek status reviews on red status projects encourage immediate corrective action (including activities performed over the weekend) to put them back on track for the next week's review cycle beginning on Monday.

Activity #5: Develop Universal and Tailored Training

As stated above, an outcome of Activity 4's deep technical review is the OPM System user guides, tutorials (if available), and vendor-provided system specifications that will form the basis and input for both technical support and end-user training.

- Determine the training strategy: classroom-based, distance learning/instructor-led online, computer-based training (CBT), self-paced via Learning Management System (LMS) or hybrid approach
- Develop requirements universal for all end users, focused on individual needs of the main functional groups represented
- Develop specially tailored training requirements for technical support staff, managers, PMO, as well as any regional variations previously identified in the planning process
- Gather templates and remaining business process documentation
- Assemble training curriculum outline and learning objectives
- Upload the training program outline and learning goals into the LMS (Learning Management System) (if available)
- Gain approval for the curriculum outline while assembling the initial drafts from universal requirements into the design of the training program (typically done in MS PowerPoint®)
- Develop the training content and incorporate additional media, videos, hands-on exercises/simulated learning using the OPM System templates
- Stage training data into a separate/replicated OPM System training environment; ensure a backup is taken to facilitate future restores of the staged data

Consider the development of separate training curricula for the previously identified functions:

- Leadership training: Developed for the senior leaders, Executive Committee, and Portfolio leaders to review project reporting and project outcome data options. Two to three hours of virtual or classroom-based training would be sufficient to complete the learning objectives, but additional time for Q&A, feedback, and follow-on planning should also be factored into the training program.
- Project Managers: The curriculum and learning objectives for this group would be the most comprehensive, including scheduling and management of portfolios, programs, and projects in the OPM System emphasis on Project Change Management and Reporting. This could potentially encompass 20 hours of classroom-based training.

- Manager Training: Curriculum developed for mid-level managers of people and support functions impacted by the ongoing Portfolio Management operations; emphasis on Change Management. Classroom-based, eight hours.
- Project Team Members: Participation would be limited to employees solely focused on completing individual tasks within a project's Work Breakdown Structure (WBS); emphasis on individual task management and recording time against the project. One to two hours virtual, online, or classroom-based, but could be more if an emphasis on Change Management is required to reduce resistance to change.

Don't make the mistake of thinking that this is a one-time training program. It has to be an ongoing training program where new projects/ programs are added to the list of active projects/programs and are available as new people move in and out of their various project roles and assignments. At the executive level, this often requires individual tutoring sessions for the new executives and follow-up sessions as individual second- and third-level managers begin to refocus their priorities away from the project at hand. Without the proper structure and culture of learning, training, and regular reinforcement, it becomes very easy for managers to become distracted by their ongoing work and lose focus on the projects and programs that are designed and chartered to secure the future of the organization.

Activity #6: Develop Measurement and Reporting Standards

The Executive Committee, with support from the PMO and local management, is responsible for development of metrics, and reporting standards and protocols to retrieve information on project/program schedules and progress to internal and external stakeholders. This would continue postimplementation of the OPM System into ongoing PMO reporting operations.

- Reporting the progress of work at project level.
- Project/program milestones and due dates; provide minimum standard framework of core milestones for all projects (e.g., completion of charter, project plan, status reporting).
- Processes and procedures to monitor and report status on tasks and milestones completed and those overdue; allow for flexibility for project optional milestones.

Audience	Project Team	Portfolio Leader	Project Sponsor	Executive Team
Metric Category	Metrics	Metrics	Metrics	Metrics
Resource Needs and Constraints	Time on task/ start-finish vs. estimate	# and % of free resources	% resource utilization	% projects increase/ decrease
Value vs. cost	# of days (tasks) past due % complete	% complete within baseline Milestones complete	$ and resources reallocated	% of portfolios/ projects on time, within budget
Portfolio health monitor	# issues resolved	# issues and risks Risk Priority Number	# of projects in green, yellow, red status	# and severity of unmitigated risks
Budget and Finance		$ currently allocated (single portfolio)	% variance $ allocated	$ currently allocated (all portfolios)
Portfolio Diversity		% of portfolio(s) in strategic alignment	# portfolios dedicated to core business needs	Ratio of portfolios on enterprise vs. short-range focus

FIGURE 4.2
Sample portfolio reporting metrics and audience.

- Ensure adherence to scheduling templates to provide standard reports throughout the OPM System.
- Listed in the matrix in Figure 4.2 is a sample of the portfolio reporting metrics many organizations choose to track, and the suggested audience for each metric.

Activity #7: Identify Risks and Technology Constraints

Each project has its own life cycle, progressing through a series of phases from initiation, through the actual work performed, and project closure. The project life cycle is categorized within five Process Groups, each a logical grouping of Project Management, various input and output data, tools, and methods:

- Process Group 1: Initiating
  Processes performed to define and authorize a new project or phase to start. Outcomes include scope and charter, overall outcome, stakeholders, project manager, and team.
  - Business case assessment, approval, and funding are external to project boundaries, although the team may be involved in the business case development.
  - The key purpose of Initiation is to align stakeholder expectations with project purpose and enroll them in the scope and objectives to illustrate how they can impact the success of the project.
  - The primary risks at this stage are lack of stakeholder alignment or misaligned objectives.
  - Additional risks include underestimating the resources required to complete the project including time, money, skills, equipment, and human resources.
- Process Group 2: Planning
  Includes processes required to scope, refine objectives, and define actions to meet project objectives.
  - The primary risks at this stage are inadequate scoping, lack of boundaries, and lack of planning for how to adapt to the inevitable project impacts which will arise.
- Process Group 3: Executing
  Includes processes performed to complete the tasks in the project plan to meet project objectives.
  - Risks at this stage are unforeseen events, or inability to adapt to changing requirements. Slippage and lack of execution/timeliness and quality of results are always potential risk areas to monitor.
  - Possibly the greatest risk is lacking of proper human resource skills when the plan calls for using them.
  - The second greatest risk is scope creep causing the project to exceed budget and schedule.
- Process Group 4: Monitoring and Controlling
  Includes ongoing "background" processes required to track, review, and regulate the project progress and success in all five process stages.
  - The primary risks at this stage are inability to manage change, evaluate, scope, and initiate changes.

- One of the biggest problems is the accuracy of the data as estimated by the people performing the tasks. In many cases, without the proper guidance and structure, people have a tendency to overcommit and underperform.
- Process Group 5: Closing
  Includes processes performed to finalize all activities and formally close out the project (or phase).
  - The primary risks at this stage are inadequately documenting the lessons learned or lack of handoffs required to ensure the ongoing success of the initiative.

The process groups and risk areas described above illustrate where key checkpoints should be included and provide guidance on where to evaluate during process checkpoints to minimize risk.

The PMO with guidance from the Executive Committee is responsible for the identification and mitigation of risks and constraints during the implementation of the OPM System and throughout the Portfolio Management life cycle. When performing this risk assessment, project managers should look to the following commonly risk-prone areas:

- Explore vulnerable areas to identify potential risks and technology constraints.
- Integrations with external systems are particularly risk prone.
- However, in many cases, the information needed to populate the OPM System is readily available in another system, and integration offers tangible benefits in time savings and elimination of data duplication and duplication of effort.
- Information may include prior project identifiers or project codes, costing and budgets, locations, resources, and other common project data elements.
- Develop mitigation action steps and solutions.
- Identify requirements for data migration/integration, including field mappings.
- Review procedures for management of baselines in the OPM System.
- Ensure security protocols to ensure baselines used to report targeted completion dates are not wantonly updated or modified.
- Projects and activities—and associated data—represent another area of potential technical constraints, which are common OPM Systems'

limited (initial) capability of summarizing detailed task/WBS data into a summary format or rolled up into project reports.

- Additional support may be required to create interim extracts or manipulate extracted data into reports to provide the needed granularity during the initial phases of the OPM System implementation.

Activity #8: Schedule and Facilitate User Acceptance Testing and End User Training

The reason why scheduling and actual facilitation of the end-user training sessions is not done until Activity #8 is because

1. the work of developing the training program is generally done concurrently with the development of the OPM System reporting methodology;
2. it also allows for the mitigation of any risks; and
3. waiting until this stage allows time to address the typical challenges of scheduling a complex training program to reach multiple locations and user populations across the OPM System's span of control.

Following are some guidelines for developing and structuring the user acceptance testing (UAT) and end-user training materials:

- Develop User Acceptance Testing (UAT) scripts based on existing/vendor provided job aids, as well as the developed OPM System end-user training materials.
- Create basic separate agendas for both UAT and training participants providing an overview of the purpose and basic outline of participant expectations and rules of conduct.
- Identify participants ideally suited to participate in UAT (potential power user, strong attention to detail, adequate computer and documentation skills, understanding of possible system errors, continuous improvement mindset).
- Publish and communicate the schedule for the OPM System UAT sessions.
- Facilitate the UAT sessions ensuring that any findings and issues are documented, and evaluate participant responses for immediate corrective action/defect/bug reporting.
- Allow time for the majority of fixes to be made and retested prior to holding the various end-user training sessions (you don't want your general user training population catching bugs!).

- Publish and communicate the schedule for available training classes.
- If possible, have an Executive Committee and/or PMO member present during each session to address any issues that may arise.
- Facilitate the training sessions and evaluate participant responses for continuous improvement.
- Document common questions and answers provided during the sessions to enhance training and support materials and develop FAQs for common scenarios.

Upon conclusion of training and UAT, be sure to review and prioritize OPM System implementation impacts and issues requiring immediate attention, versus those scheduled for updates in a later release.

Activity #9: Develop Project/Portfolio Security and Data Integrity Procedures

At this stage of the OPM System implementation, Information Security is engaged to ensure the integrity of project and portfolio data, including the development of user access controls (UAC) to regulate access for the general user population and restrict access to administrative system functions.

- Establish the basic database structure that will organize portfolios, programs, and projects
- Define and assign the access rights/permissions for the general user population
- Review processes, policies, and procedures to ensure security and data integrity protocols align with the intention of these previously documented OPM System artifacts, including system requirements and specifications, also possibly specified within the original Technology/Information Security & Disaster Recovery Plans for the project
- Establish ongoing regular reviews to ensure the ongoing security and data integrity of the OPM System

Particular attention should be paid to the protection of the information from data and security breaches. Much of the information included in the projects relate to programs, processes, and products that are designed to give your organization a competitive advantage. Hackers obtaining this information and selling it to competitors around the world presents a major problem in today's networked global marketplace. Organizations proceed at their own peril until they take the steps to ensure their information is secure from outside illegitimate access to it. The 2014 cyber attack on Sony Pictures Entertainment's network of systems, servers, and databases is an excellent example of how unmitigated risks can have a major negative impact on an organization's ability to perform.

In December 2014, a class action suit against Sony in California was filed in the federal court system. If the data breach itself wasn't warning enough, this lawsuit should serve as a call-to-action for companies to finally put the appropriate security measures in place to prevent data breach, loss of competitive advantage, and the resulting fallout of public perception and lost customers. The complaint outlines several measures that organizations must consider to safeguard themselves against security breaches and resulting legal fallout and public backlash. Paragraph two of the complaint outlines the path forward for organizations seeking to failsafe their networks and reputations against similar damage (Figure 4.3).

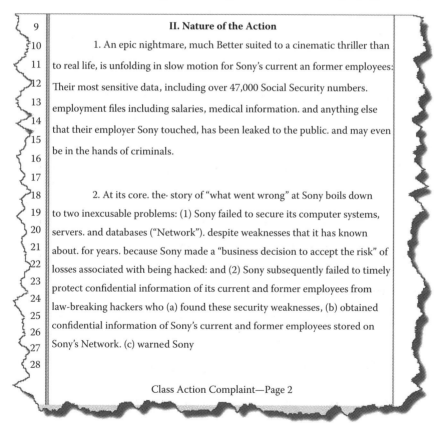

II. Nature of the Action

1. An epic nightmare, much Better suited to a cinematic thriller than to real life, is unfolding in slow motion for Sony's current an former employees: Their most sensitive data, including over 47,000 Social Security numbers. employment files including salaries, medical information. and anything else that their employer Sony touched, has been leaked to the public. and may even be in the hands of criminals.

2. At its core. the· story of "what went wrong" at Sony boils down to two inexcusable problems: (1) Sony failed to secure its computer systems, servers. and databases ("Network"). despite weaknesses that it has known about. for years. because Sony made a "business decision to accept the risk" of losses associated with being hacked: and (2) Sony subsequently failed to timely protect confidential information of its current and former employees from law-breaking hackers who (a) found these security weaknesses, (b) obtained confidential information of Sony's current and former employees stored on Sony's Network. (c) warned Sony

Class Action Complaint—Page 2

FIGURE 4.3
Class action complaint.

An organization's OPM System should fall under a larger umbrella security policy, intended to address these types of core structural inefficiencies:

1. Secure the systems, servers, and databases against known and unknown security threats.
2. Protect the confidential information of the organizations plans, stakeholders, management information, and that of current and former employees.

The suit makes clear that organizations, including Sony, owe a legal duty to maintain reasonable and adequate security measures to secure their networks, and the data that resides on them (Figure 4.4).

We can all craft a plan for how to proceed from the allegedly deficient actions in this case:

- Consider creating and documenting an Information Security Plan (ISP), with provisions for both prevention (identifying and resolving present security vulnerabilities) and what to do, and how to recover and respond (and who to contact) in the event of a security breach.
- Design and implement adequate firewalls and computer networks.
- Ensure measures for data encryption are in place (including randomization if required).

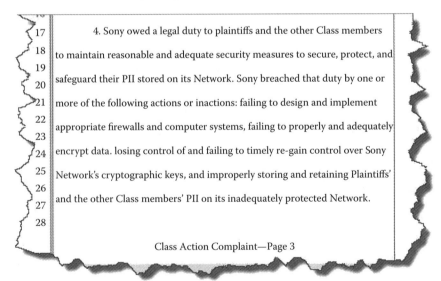

17
18
19
20
21
22
23
24
25
26
27
28

4. Sony owed a legal duty to plaintiffs and the other Class members to maintain reasonable and adequate security measures to secure, protect, and safeguard their PII stored on its Network. Sony breached that duty by one or more of the following actions or inactions: failing to design and implement appropriate firewalls and computer systems, failing to properly and adequately encrypt data. losing control of and failing to timely re-gain control over Sony Network's cryptographic keys, and improperly storing and retaining Plaintiffs' and the other Class members' PII on its inadequately protected Network.

Class Action Complaint—Page 3

FIGURE 4.4
Class action complaint.

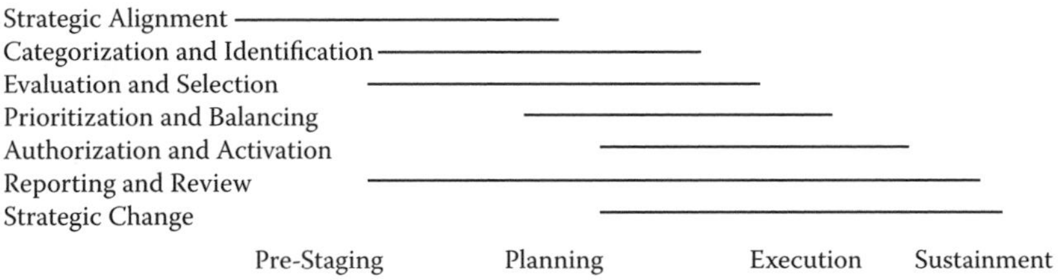

Strategic Alignment ———————————
Categorization and Identification ———————————
Evaluation and Selection   ———————————
Prioritization and Balancing   ———————————
Authorization and Activation   ———————————
Reporting and Review   ———————————————————
Strategic Change   ———————————————

         Pre-Staging   Planning   Execution   Sustainment

FIGURE 4.5
Sample Portfolio Management System implementation phases.

Activity #10: Implement the OPM System and Report Progress

- Review results from UAT and pilot deployments.
- Incorporate findings into the OPM System Implementation Plan.
- Activate change management protocols for communication of the pending implementation (see Chapter 5).
- Implement the OPM System (Figure 4.5). Typically there are not defined lines for when one phase ends and another begins. This demonstrates the concurrent and overlapping nature of the implementation phases.
- Monitor the results in real-time using intraday communication and vertical reporting.
- Report progress to the Executive Committee, organizational leadership, and stakeholders.
- Schedule follow-up meetings to review deployment findings and process lessons learned.
- Ensure migration of program/project data and schedule sunset of legacy systems (if applicable).
- Audit ongoing utilization of schedules, templates, and Portfolio Management System procedures.

SUMMARY

In this chapter, we covered the 10 suggested activities for successful implementation of the OPM System. As with Portfolio Management itself, the essential first activity requires involving the senior leadership of the organization in a review and planning for communicating a clear vision of the organization's strategic goals and objectives. Once the leaders themselves are aligned behind the vision, the Portfolio Steering Committee will typically marshal additional (temporary) resources from HR, communications, and the PMO to articulate and communicate the change agenda. This includes the initiative's goals, objectives, benefits, risks, rewards, and challenges, while clarifying the roles and responsibilities and involving impacted associates and potential change agents in the planning process. With guidance from the Portfolio Steering Committee and Project Team manager, the assembled resources also are involved in identifying impacted business processes and ensuring that required resources, such as IT support resources, are provided adequate notice regarding their planned involvement during the planning and implementation phases of the project.

In order to enact the change agenda and ensure the required end-user adoption, careful consideration is given to develop custom-tailored training and communications, as well as robust measurement and reporting standards, including information dashboards for executive-level review of portfolio status and performance.

In order to fail-safe the implementation of the OPM system, a risk analysis is performed to forecast any potential risks, technology constraints, or potential environmental factors that may arise during the course of the implementation.

Prior to implementation, the change management process begins in earnest as the organization schedules and facilitates user acceptance testing (UAT) and end-user training sessions geared toward providing the crucial user experience feedback on usability and performance while providing for a valuable last opportunity to bug check the system before wide-scale deployment.

In addition to the supplemental/nonfunctional usability and performance testing, development of project/portfolio security and data integrity procedures provides additional assurances that the organization has done everything in its power (balancing risk and prevention costs) to safeguard the organization's vital private data against emerging cyber security threats.

Finally, the Program Management Office (PMO), IT, and Project Teams have learned everything they can from the pilots and are ready to implement the OPM System and report on both the progress of the implementation, as well as establish the time recording, feedback, and reporting structure for the individual portfolios, programs, and projects under the OPM System's span of control.

REFERENCE

1. Yaskawa Electric Corporation. 2015. Vision 2015/Direction for FY2015. Online at: http://www.yaskawa.co.jp/en/vision/index.html

5

Phase IV: Change Management—
Practical Applications within OPM

INTRODUCTION

Organizational Portfolio Management (OPM) involves identifying and aligning the organization's priorities and establishing governance and a framework for performance management (and continuous improvement, if warranted). Our research shows how project/program managers, Portfolio leaders, and their organizations can move beyond installation (of the OPM System) into the realm of realization and repeatability. Integrating these seven essential Project Change Management practices within Project/Program and Portfolio Management applications can help you realize results consistently by applying the Portfolio Change Management Model (Figure 5.1).

ACTIVITIES FOR PHASE IV

Phase IV (*Practical Applications of Project Change Management within OPM*) is made up of the following seven activities (Figure 5.1):

Activity #1: Start at the top
Activity #2: Create a Portfolio Enrollment and Management Plan
Activity #3: Communicate the rewards, challenges, risk, and consequences
Activity #4: Build capacity within the organization
Activity #5: Integrate risk mitigation and project planning
Activity #6: Plan for sustained results
Activity #7: Standardize the Portfolio Change Management approach

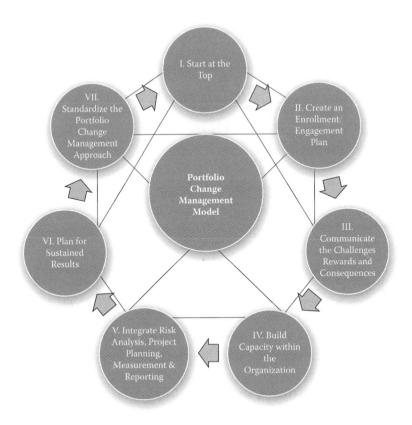

FIGURE 5.1
Portfolio Change Management Model.

Activity #1: Start at the Top

In order to attain the stated goals of the Organizational Portfolio, individual and collective behavioral change must be attained. If behaviors within the organization are to change, accountability needs to begin and end with the Project Team manager and the Executive Committee. Successful Organizational Portfolios seek to accomplish 95 to 100 percent of the stated goals. In order to accomplish this over the long haul of a portfolio life cycle, senior executives must be enrolled first and commit to not only the goals of the program, but must embrace and embody the goals of the change effort by seeking to first attain alignment amongst themselves, so they are all on the same page before they go out and attempt to enroll the rest of the organization in the Change Agenda.

Think of the Change Agenda as a living document; an actual working agenda that outlines and communicates planned events, which gets disseminated as part of the Communication and Enrollment Plan. Much like the purpose, agenda, and limit (PAL) used for effective meeting agendas, the Change Agenda establishes the purpose, components, and scope of the change effort, and is distributed with the schedule of project kickoff meetings, enrollment/training sessions, and conference calls held to explain the purpose, scope, and milestones of the project.

To ensure a deep alignment of the Change Agenda, begin by establishing alignment on the senior team with the business model and strategic plan. Then seek to enroll the people who are driving and being impacted by the change.

Once the Portfolio Development Team attains consensus and alignment on the direction of the programs to be active in the current project cycle, it becomes a powerful force as the Portfolio Development leaders ensure that alignment occurs below them as well in parallel interfaces with other areas of the organization. Portfolio Development leaders also can ensure mission and resource alignment across their portfolio(s) of sometimes diverse and distinct programs and projects, which otherwise might not interface with one another. When enforcing the consequences of change action, it becomes important for the Executive Committee to back one another as well as leverage interdepartmental cooperation. There is no better way to ensure adoption rates than an actively engaged and committed leader placing a priority on the Change Agenda within their organization.

With the appropriate priorities and potential positive (and negative) impacts understood, change agents will emerge as the organization takes stock of strengths, capabilities, and resources that can play a key role in the change effort. Starting at the top becomes evident when the senior executives back up their talk with sustained positive action in promotion of the Change Agenda, visibly and audibly supporting the goals of the Change Agenda when communicating with their staff. This is essential for Portfolio leaders, program managers, and project managers when interacting cross-functionally with other organizational units or departments that may be contributing to or impacted by their portfolio. The Executive Committee's key responsibilities are to report on the overall success of their initiatives to shareholders, and ensure allocation of shared resources across multiple projects, programs, and portfolios. At the same time, they are working to resolve multiple cross-functional issues, remove barriers, and eliminate resistance to change by promoting and enrolling others in the change effort.

One way to assess the impact of the change effort is through the organization's formal and informal social media channels, as well as through the regular business of project and program review conducted by the Project Management Office (PMO) and Portfolio Development Team. The goal is to find a way for the employees, senior staff, and potentially customers to interact with senior executives, not only to demonstrate their involvement but out of a genuine active interest in the change effort borne of their own efforts and engagement.

Activity #2: Create a Portfolio Enrollment and Management Plan

The Change Agenda sets the parameters and schedule for the initial enrollment/communication and planning sessions. The purpose of these initial enrollment/communication and planning sessions is to establish a two-way dialog, to engage people in the concept and benefits (and challenges) while at the same time enroll them in the effort, explain their roles, and show them how they will be active participants in crafting the plan for change.

Creating and communicating a crystal-clear Change Agenda sets the foundation for a successful portfolio of programs and projects, be it the implementation of a new or updated system, a departmental organization, an organization-wide restructuring, or the creation of a new program, product, or service line.

The Portfolio Development Team is also responsible for creating Enrollment Plans (see Chapter 3, Activity 4), which further outline the desired behaviors, actions, accountability/ownership, and targeted completion date.

Each project should have an Enrollment Plan to drive alignment from the top down through the middle layers of the organization, to ensure the employees being impacted by the change are aware of the purposes and consequences of the change. Informed employees are empowered to help drive change or simply work collaboratively and individually within the project to best assist in the implementation and helping to close the gap between mere installation and wider adoption. One practical way to track enrollment is through end-user adoption and usage rates. Your systems and database administrators should be able to provide transaction counts per day, as well as the number of unique user sessions within the system on a daily basis. The goal in calculating adoption/usage rates is to compare the number of registered users with those actually using it on a daily or weekly basis.

As a key component of the Stakeholder Management Plan, when creating the Enrollment Plan and Change Agenda, consider these five essential factors for sustainable change:

1. Stakeholders know the purpose of the project and why it is important for their team and the organization.
2. Stakeholders receive communication outlining the scope and key milestones of the project.
3. Stakeholders are enrolled in the essential elements of the project plan relevant to their roles.
4. All stakeholders understand the critical success factors (CSFs) and potential areas of risk, and participate in taking reasonable actions assigned to ensure success and mitigate risks (as delegated in the Enrollment/Engagement Plan).
5. Stakeholders are assigned responsibility for the portions of the change related to them. This ensures direct accountability throughout the project life cycle.

Every person in the organization needs to have his or her role clearly defined no matter how great or how small—from project and program managers, facilitators, team leaders, team members, subject matter experts, end users, and support staff. Everyone in the organization must be enrolled in the change effort.

Furthermore, employees must have an incentive to change, a motivation for action, and an understanding of the risks and consequences of both failure and success. To ensure alignment with the Change Agenda as part of a Stakeholder Management Plan, the enrollment plan should contain a resource matrix consisting of the resources driving and impacted by the change to ensure they are adequately motivated and inspired to act.

The Portfolio Resource Enrollment/Alignment Plan establishes a resource matrix consisting of the impacted/involved resources and the percent of their time dedicated to the change effort. The resources should consist of change sponsors, change agents, and acolytes as well as devil's advocates and potential resisters (those who may have previously expressed reservations or exhibited lack of commitment to the change effort). Because the senior leaders responsible for OPM are typically involved in multiple programs, projects, and initiatives, their responsibility extends from allocating resources to working to remove barriers to vertically promote the change effort, as well as roll up reporting to the Portfolio Steering Committee on the overall success of their initiatives.

Activity #3: Communicate the Rewards, Challenges, Risk, and Consequences

Once alignment has been attained among the Executive Committee and within the organization's strategic objectives, communication plans can be created to begin the process of enrolling the organization. In addition to the basic who, what, when, and how, the communication plan should clearly communicate the shared vision and purpose for the change, by honestly outlining the opportunity (or problem), its associated challenges, and the rewards that will come about as the project goals are realized. When dealing with consequences (and they must be dealt with up front), the communication needs to strike a balance of both positive and negative consequences based on objective measures.

The Communications Plan is vital in setting the stage for ongoing and constant communication throughout the project life cycle, and allowing alignment to occur, which cascades into individual and collective enrollment. Lack of alignment between business units responsible for strategy execution is one of the biggest causes of project failures. Inadequate information sharing causes that lack of alignment. When you begin at the top, create an enrollment plan, shift people's paradigms toward alignment, and have leadership teams communicate and demonstrate what it will take for everyone (themselves included) collectively and personally, you create dynamic vertical alignment for the change. Part of the reward will be the organization's investment in the development of new/enhanced skill sets and the consequences can be enforced by tying adherence and adoption directly to individual development plans (IDPs).

Senior leaders must drive communications and consequences aligned with the change management plan. The communication plan then becomes more than a document of intent to explain the project to the rest of the organization. It goes beyond that by articulating that a deep alignment with the organization's strategic plan has already begun and helps the audience understand how they can get onboard—to become change agents—in order to go out and enroll their people. By clearly articulating the purpose of the change, the consequences, and the next steps within the enrollment plan (see above), the stage has been set for the organization to deal with and ultimately embrace change. Helping overcome resistance to change is the fact that the current state has gotten us to where we are today, and, if the current state is a burning platform, everything the organization has been working toward could be at risk.

Articulate a clear and compelling vision for where you are heading, but be realistic. Project Team managers need to be aligned and committed to the vision, understand how disruptive the change is going to be, and balance the ambition with the organization's capacity to absorb that change.

Activity #4: Build Capacity within the Organization

Resilient organizations have the ability to absorb high levels of disrupted change while displaying minimal dysfunctional behaviors. Resiliency requires capacity for change; it requires an organization focused on common objectives and is observed through individuals and synergistic groups having positive attitudes leading to results and a proactive approach to change—one that is both flexible and organized.

Many organizations go through the motions in predictable patterns and routines, including those that directly contribute to what made the organization great in the past. However, what made your organization great in the past may not be the case today, and may not be nearly strong enough amid increasing global competition to sustain investor demands for increasingly stronger profits and performance.

To break dependence on the status quo, successful Portfolio leaders highlight the gaps between current and desired performance. Capacity building is further enhanced by creating an appealing vision of the future state, and fostering confidence that the future state can be achieved. Building capacity starts with strategic project planning, while training enhances motivation for change. When people are measured by the success of desired project outcomes, and their adaptability to change becomes part of the regular performance review cycle, it tends to help drive adoption rates.

Concurrently, address the situations causing high stress levels within an organization that can cause the organization, and projects that operate within them, to become dysfunctional. Managing the stress that the organization is subjected to will eliminate one of the major root causes of project failure, and minimize negative performance resulting from capacity overload.

Identify the few critical behaviors required to drive results; shift and reinforce behaviors by changing consequences and measuring progress.

Activity #5: Integrate Risk Mitigation and Project Planning

With the Portfolio Development Team and Portfolio Team managers aligned, committed to and embracing the change, the change agents are typically the ones who know best how the change may impact their people and processes, and plan accordingly to mitigate the risks associated with the change. A balance needs to be struck between the desire for change and the organization's ability to embrace any fundamental changes.

In OPM Systems, projects must be time bound to avoid impacting the milestones and resources of other programs and projects that may be intertwined. One of the goals of many projects is completion within a specific timeframe, which begins the negotiation process for allocation of resources. The greatest overall risk that portfolios face is an inadequate initial alignment, which causes the programs and projects to be misaligned through the portfolio life cycle. Programs face the risk of competing or inadequate resources and budgetary constraints, while project risks typically center on missed deliverables and scope creep, both of these time-constrained. Doing an effective job of risk mitigation up front can save a tremendous amount of time down the road, but you cannot always mitigate everything.

The prioritization of change-related risks, ranging from inadequate adoption, knowledge transfer failure, all the way to employee exodus and catastrophic system failure, is critical at this juncture during inception of the OPM System.

Risk is one of several factors the Executive Committee uses to evaluate the impacts of sweeping change brought about the realization of implementation of 95 to 100 percent of a portfolio's stated objectives.

When analyzing potential impacts and seeking to mitigate unwarranted levels of strain within the organization, factors for consideration include urgency, risk, benefits, time, costs, and impact. When assessing impact, consider impacts on the customer (beneficial) as well as on/within the organization (less desirable because this represents additional resource requirements).

Every portfolio, program, and project must have a risk analysis prepared for it. There are several methods to assess and mitigate risk, depending on the nature of the system and the type of data available. Following is a partial list of some of the more commonly used risk analysis methods available today:

- The Risk Probability and Impact Matrix assesses potential fail points and risks for their likelihood of occurrence and the impact they may have on the portfolio/program/project.
- Monte Carlo simulation software applications (or spreadsheets) mathematically predict failures or overruns in complex, dynamically changing systems and environments.
- The Failure Modes Effects Analysis (FMEA) method is effective for group brainstorming on potential failures by assigning a Risk Priority Number (RPN) to potential impacts with higher likelihood and severity, compounded by lower detectability. For example, a project may have a risk associated with user adoption. With mid-level severity of 5/10, likelihood of occurrence at 5/10, and a 50-percent chance of detecting and preventing the issue, the RPN would be 125 (derived from severity × occurrence × detection or 5 × 5 × 5 = 125).
- The Harmful/Useful (HU) diagram method is used to analyze a particular product and/or solution to define harmful and useful aspects of the proposed product or solution. Each of the harmful aspects represents a risk or negative aspect to the proposed action. Analyzing these harmful aspects and implementing useful aspects to offset them is another way of identifying risks and mitigating their impact.

Whichever method is used for the risk analysis, the various risks should be prioritized, with actions identified to mitigate higher priority risks. These risk mitigation actions should be integrated into the overall project plan, the measurement plan, and portfolio, program, and project reporting to assess shared impacts. Portfolio, program, and project reporting becomes part of the performance feedback channels which Portfolio leaders utilize to promote change and report to senior leadership on overall performance.

The results of the risk analysis, including any tasks tabled for a later phase and especially actions identified to mitigate real or potential risks, must be incorporated into the Portfolio Risk Register (the active log of recorded risks and mitigations), and into the program and project plans, stakeholder management plans, enrollment plans, training plans, and communication plans. As with all elements of the plan associated with critical success factors, these CSFs (derived from portfolio, program, and project goals and objectives) collectively will become the basis for real-time measurement and reporting. These feed into program, project, and process quality control measures that will continue after the goals of the portfolio have been completed and incorporated into final project reporting recommendations to ensure ongoing sustainability.

Finally, during the execution phase of the portfolio (and all of its associated program, project, and related plans), the initial goals established (with realistic measurable targets) become the basis for ongoing measurement. The initial goals also will be reported again in the final closure reports including status on any outstanding risks, tasks, milestones, and deliverables.

Activity #6: Plan for Sustained Results

Through practice and iterations of change, the investment in change will be realized once the initial successes can be replicated beyond implementation into ongoing operations and controls designed to ensure sustainability.

As additional capacities become available and the organization becomes accustomed to adapting to change, it yields a well of resources that can be converted into sustaining prior gains, investment in innovation, new capacities, product, and service lines. This leads to competitive market advantage. Building further capacity for additional iterations of change and innovation through successful portfolio management is the goal.

Management's ability to influence the organization's direction with successful change initiatives drives perceptions within the organization. The focused Portfolio Change Management approach determines and aligns employee perception with the organization's strategy and vision of itself in the future, then reinforces these images throughout the transformation process.

By following the Portfolio Change Management Model, plans for near-term enrollment and long-term sustained results are established to address real cultural and perceived barriers to change. By planning to address the human element of change, organizational resistance to change is reduced. This leads to greater stability in the short term, and sustainability in the long term (as long as the efforts to reduce resistance to change are ongoing), and the OPM System is continually fine-tuned to process changing organizational strategic objectives and adaptive learning from prior portfolio program and project results.

Activity #7: Standardize the Portfolio Change Management Approach

As the name implies, the Portfolio Change Management approach starts at the top by ensuring only portfolios, programs, and projects that align directly with the organization's top strategic objectives are approved. With these strategic objectives as the framework, development of the portfolio planning, scoping, and initial objectives are fashioned after the organization's goals and objectives.

Focusing and standardizing organizational change approaches across the Organizational Portfolio over a sustained period of time is crucial to creating an organizational culture that accepts change as part of the organization's normal operations.

The value in taking a Portfolio Change Management approach to change is that it directly links the change agenda back to the organization's strategic plan objectives. Furthermore, a portfolio's collective components (programs, projects, or even subportfolios) are a direct reflection of the organizational strategy and objectives. Portfolio management involves identifying and aligning the organization's priorities, establishing governance and a framework for performance management (and continuous improvement if warranted). When taking a Portfolio Change Management approach, the organization's projects and programs roll up delivering comprehensive portfolio reporting and assessment of value, cost/benefit, and return on investment (ROI), while allowing a broad vantage point for evaluating risk and the allocation of resources across the system.

A leading provider of information analytics, Forrester Research reports that OPM tools improve business outcomes by 10 to 25 percent (varying by organization) on measurable factors, such as:

- Process improvements and reporting automation resulting in productivity enhancements
- Improved processes for OPM resulting in improved performance against strategic organizational objectives
- Cost reductions (from implementation of Software-as-a-Service/SaaS OPM solutions) and avoidance (from reduced reliance on hardware and maintenance costs)
- Improved project completion success rates, from improved oversight and collaboration

Best practices in OPM benefit organizations by ensuring standardization of the following practices, typically the province of the organization's PMO, are applied to the change agenda:

- Providing an infrastructure for the management of projects, programs, and portfolios and the execution of individual change agendas
- Supporting review and evaluation of new initiative requests, facilitating prioritization and authorization of new projects, and allocating resources to affect change in alignment with organizational strategy and objectives
- Providing project and program progress reporting of critical success factor metrics, resources, expenditures, defects, and associated corrective actions to the respective Program Management Office and Change Control Committees
- Negotiating and coordinating resources between projects, programs, or other portfolios
- Assisting with risk identification and mitigation
- Communicating risks and issues related to ongoing initiatives across the portfolio of projects, programs, and subportfolios
- Monitoring compliance to PMO policies and ensuring ongoing alignment with the organization's strategic objectives
- Mentoring change agents while developing and delivering training in process, project, and change management tools and techniques
- Providing knowledge management resources, and archival services, including collection and propagation of lessons learned

Once the objectives for change have been attained and reported, it's time to close out the project. This last step is very often overlooked because a bit of change exhaustion may have set in during the course of the initiative. Understand that this dynamic will be present as your project winds down and energy naturally shifts elsewhere; plan up front on documenting the lessons learned at this stage (transition or closure).

Beyond this, most of the real change takes place after the project is over. By the time the project has been implemented and integrated within the normal day-to-day business operations, the project team has been disbanded. New (and hopefully value adding) user habits are formed during this early postimplementation stage. The good habits required to sustain the changes only become ingrained after months of continued application under normal/optimal circumstances.

SUMMARY

Experts in the field of Change Management universally agree that the success or failure of a change initiative is not just about initiating, planning, monitoring, executing, and evaluating the project that will drive the change. The organization must transition to a culture that embraces change, ensures stakeholder buy-in, and engages executive sponsors to align and support the change before, during, and after its implementation. By following the Portfolio Change Management Model outlined in the chapter, organizations will be better positioned to maximize the return on investment of time and resources invested in their portfolios, programs, projects, and people. This results in organizational transformation by enrolling and involving impacted stakeholders in working toward solutions directly aligned with the organization's strategic objectives.

Applying this Change Management-focused approach to Portfolio Management directly links the change agenda back to the organization's strategic plan results and objectives. Furthermore, a portfolio's collective components (programs, projects, or even subportfolios) are a direct reflection of the organizational strategy and objectives. When taking a portfolio management focused approach to Change Management, the organization's projects and programs roll up delivering comprehensive portfolio reporting and assessment of value, cost/benefit, and return on investment (ROI), while allowing a broad vantage point for evaluating risk and the allocation of resources across the system.

Closing/Summary of Chapters 1 to 5

Portfolio/program and project managers must adapt to ever-changing customer requirements that necessitate flexible and adaptive project schedules. Likewise, the Organizational Portfolio Management (OPM) System also must be built to adapt to decentralized organization structures and changing stakeholder requirements, while also enhancing adherence to OPM best practices in meeting the organization's strategic goals and objectives.

Involving the organization's Portfolio Development Team, Program Management Office (PMO), and impacted stakeholders in the planning process ensures successful portfolio, program, and project outcomes.

Planning, measuring, and reporting on the common OPM System critical success factors helps align the system with the organization's strategic goals and objectives, while actively and realistically communicating the purpose, benefits, and challenges of the upcoming initiatives.

Involving the impacted stakeholders includes involving and providing for the mobilization and support of the organization's Information Technology (IT) security and support networks. Human Resources becomes the focus mid-deployment through the development of universal training tailored to specific geographic and diverse demographic functional groups.

Reporting becomes the crucial bidirectional communication tool during the implementation of the OPM System and beyond into regular Program Management Office and OPM project execution.

Robust User Acceptance Testing is conducted to fail-safe the implementation. End-user training involves the organization in ongoing learning and development opportunities in open review of the OPM System. Ongoing feedback into the PMO further enhances the system. Finally, with the system adequately tested, security protocols in place, and all major deployment iterations complete, the organization begins actively utilizing the OPM System and reporting on the portfolios, programs, and projects' progress toward the advancement of the organization's strategic goals and objectives.

At first glance, an OPM system may look like increased bureaucracy laid on top of an already slow and costly project implementation model. However, when it results in a 20-percent increase in the percentage of projects that are successful, it really turns out to be one of the best resources an organization can invest in to ensure profitability and long term sustainable results.

H. James Harrington

Appendix A: Project and OPM Definitions

The following Project and Portfolio Management definitions have been excerpted and expanded on from the Project Management Institute's (PMI's) *Guide to the Project Management Body of Knowledge (PMBOK Guide)*.[1]

- **Adaptive:** Product developed iteratively, detailed scope defined for each iteration as it begins.
- **Adaptive Life Cycle:** For example, Agile Methods or Change-Driven, respond to lots of change and stakeholder involvement. Adaptive also interactive and incremental, but iterations have fixed costs and time (2–4 weeks), performing multiple processes in each iteration. Early iterations focus on planning, with particular emphasis on:
  - Scope decomposed into requirements and WBS (work breakdown structure): product backlog.
  - Each iteration begins with prioritizing which high-priority items can be delivered and ends with a delivered product for the customer to review; should have finished, complete, usable features.
  - Sponsor and/or customer representative provides feedback on deliverables and status of backlog items.
  - Preferred in rapidly changing environments, when scope and requirements difficult to obtain.
- **Business Case:** An evaluation of the potential impact a problem or opportunity has on the organization to determine if it is worthwhile investing the resources to correct the problem or take advantage of the opportunity. An example of the results of the business case analysis of the software upgrade could be that it would improve the software's performance as stated in the value proposition, but (a) it would decrease overall customer satisfaction by an estimated three percentage points, (b) require 5 percent more task processing time, and (c) reduce system maintenance costs only $800 a year. As a result the business case did not recommend including the project in the portfolio of active programs. Often the business case is prepared

by an independent group, thereby, giving a fresh unbiased analysis of the benefits and costs related to completing the project or program.

- **Business Objectives:** Business objectives are used to define what the organization wishes to accomplish, often over the next 5 to 10 years.
- **Business Plan:** A formal statement of a set of business goals, the reason they are believed to be obtainable, and the plan for reaching these goals. It also contains background information about the organization and/or services that the organization provides as viewed by the outside world.
- **Business Value:** Entire value of the business; total sum of tangible (assets, fixtures, equity, utility) and intangible elements (goodwill, recognition, public benefit, trademarks): short-, medium-, or long–term.
- **Change Agenda:** A literal agenda of events that gets distributed as part of the Enrollment Plan to all prospective change agents (CAs). It sets the stage for meetings or conference calls held with CAs to explain the purpose, scope, and milestones of the project; engage them in a dialog to enroll them in the effort; explain their roles; and show them how they will be active participants in crafting the plan for change.
- **Change Control Board:** In Information Technology and software development, a Change Control Board (CCB) or Change Control Committee is a committee that makes decisions regarding whether or not proposed changes to a project should be implemented, evaluating potential impacts to stakeholder groups or other systems. In many cases, the CCB also reviews results of implemented changes, and evaluates corrective actions in the event planned changes have any adverse impacts.
- **Closing:** Lessons learned, final project audits, project evaluations, product validations, acceptance criteria.
- **Common Experiences:** Shared mission, vision, values (MVV), beliefs; regulations, process/policy/procedure (PPP); common recognition and reward system; risk tolerance; view of leadership/authority; code of conduct, work ethic, hours; operating environments.
- **Critical Success Factors:** These are the key things that the organization must do extremely well to overcome today's problems and the roadblocks to meeting the vision statements.
- **Effective:** An output from a process or sub-process that meets the needs and expectations of end user of the output. Effectiveness is having the right output at the right place at the right time at the right price.

- **Executing, Monitoring, and Controlling:** Change control procedures; financial control procedures (time reporting, expense and disbursement reviews, accounting codes, standard contract provisions); issue and defect management procedures for identification, action item tracking, and resolution; communication requirements; prioritization, approval and issuing work authorizations, risk control procedures, probability, and impact matrix; standard guidelines, work instructions, proposal, and performance measurement criteria.
- **Initiating and Planning:** Guidelines, criteria for using organization P&P (policies and procedures) on the project; organization standards/policies (HR, health and safety, ethics, PM policies); product, project life cycles; quality P&Ps (audits, checklists standard process definitions).
- **Initiating Sponsor:** Individual/group who has the power to initiate and legitimize the change for all of the affected individuals.
- **Iterative and Incremental:** Project phases/iterations repeat project activities as the team's knowledge and understanding of the product increases. Iterations develop the product in repeated cycles. Increments add to the functionality in succession. All PM Groups performed.
  - Deliverable(s) produced at the end of each iteration, each iteration incrementally building the deliverables until the exit criteria for the phase are met. The project team processes feedback.
  - High-level vision prepared, but detailed scope is elaborated one iteration at a time, planning for the next as work is being done on the current iteration (managing and confining the scope).
  - Preferred for projects with changing objectives and scope, to reduce complexity, or when partial delivery is beneficial and does not impact the final outcome deliverable(s). Reduces risk on large, complex projects by incorporating lessons learned between iterations.
- **Knowledge Base:** Including configuration management, versions, and baselines; financial (hours, costs, budgets, overruns); lessons learned; issue/defect databases; process data; prior project files.
- **Mission Statement:** The stated reason for the existence of the organization. It is usually prepared by the CEO and key members of the executive team and succinctly states what they will achieve or accomplish. It typically is changed only when the organization decides to pursue a completely new market.
- **Ongoing work:** A repetitive process following existing procedures.

- **Operations:** Ongoing work, production of goods and services. Generally out of scope from a project/program/portfolio management standpoint. Operational stakeholders should be added to the stakeholder register and their influence (pro or con) addressed as a Risk.
- **Organization:** any group of people who work together to produce an output. It can be a team, a department, a division, or the total company.
- **Organizational Goals:** The organization's goals document the desired, quantified, and measurable results that the organization wants to accomplish in a set period of time to support its business objectives. (For example, increase sales at a minimum rate of 12 percent per year for the next 10 years with an overall average annual growth rate of 13 percent). Goals should be specific rather than general so that there is no ambiguity.
- **Organizational Master Plan:** The combination and alignment of an organization's Business Plan, Strategic Business Plan, Combined Performance Acceleration Management (PAM) Plan, and Annual Operating Plan.
- **Organizational planning:** Impacts projects; prioritization based on risk, funding, impact on strategic plan objectives.
- **Organizational Portfolio Management (OPM):** Improves organization capability, linking Project/Program/Portfolio Management with organizational facilitators—structural, cultural, technological, HR practices—to support strategic goals. To apply this methodology, organizations must first measure capabilities, then plan and implement improvements.
- **Organizational Process Assets:** "Plans, policies, procedures and processes, and knowledge bases specific to or used by the performing organization." Any artifact, practice, or knowledge used on the project policies, procedures, and processes.
- **Organizational Project Management (OPM):** A strategy execution framework utilizing OPM with organization-enabling practices to execute the strategic plan, improve performance and results. Competitive advantage **OPM** improves organization capability linking Project/Program/Portfolio Management with organizational facilitators—structural, cultural, technological, HR practices—to support strategic goals. To apply this methodology, organizations must first measure capabilities, then plan and implement improvements.

- **Organizations:** "Systematic arrangements of entities (people, departments) aimed at accomplishing a purpose, which may involve undertaking projects."
- **Policy:** A principle or rule to guide decisions and achieve rational outcomes. A policy is an intent to govern, and is implemented as a procedure. Policies are generally adopted by the Board of Directors or senior governance body within an organization, whereas, procedures are developed and adopted by senior managers. Policies can assist in both subjective and objective decision making.
- **Portfolio:** A centralized collection of independent projects or programs that are grouped together to facilitate their prioritization, effective management, and resource optimization in order to meet strategic organizational objectives.
- **Portfolio Components:** Constituent programs, projects, and other related work. Status reports, lessons learned, and change requests roll up to the portfolio.
- **Portfolio Development Team:** Review/prioritize projects and programs for resource allocation, aligned to organizational strategies.
- **Portfolio Leader:** A senior project manager qualified and appointed to manage multiple concurrent and interdependent subportfolios, programs, and projects. Portfolio leader roles are typically awarded to program managers with years of demonstrated success organizing and managing programs with multimillion-dollar budgets allocated from (and aligned to) the organization's key Strategic Objectives. (See also Portfolio Management.)
- **Portfolio Management:** Aligns organizational strategy by prioritizing programs and projects, prioritizing work, and allocating resources. It is the "centralized management of one or more portfolios to achieve strategic objectives."
- **Portfolio Manager:** A senior project manager qualified and appointed to manage multiple concurrent and interdependent subportfolios, programs, and projects. Portfolio manager roles are typically awarded to program managers with years of demonstrated success organizing and managing programs with multimillion-dollar budgets allocated from (and aligned to) the organization's key Strategic Objectives. (See also Portfolio Management.)
- **Predictive:** Plan-driven product and deliverables defined at the beginning and scope carefully managed.

- **Predictive Life Cycle:** For example, requirements, feasibility, planning, design, construct, test, turnover; preferred when the outcome is well understood, a base of industry practice exists, or the outcome needs to be delivered in full in order to have value to stakeholders:
  - Rolling wave planning: General macro plan created, then more detailed planning as resources need to be assigned for specific phases.
- **Process:** "A set of interrelated actions and activities performed to create a prespecified product, service, or result." Each process is comprised of inputs, outputs, tools, and techniques, with constraints (environmental factors), guidance, and criteria (organizational process assets) taken into consideration.
  - Select appropriate processes to meet project objectives, adapt a defined approach to meet requirements, communicate/engage stakeholders, meet needs, balance constraints.
  - Project Management processes: Ensure flow throughout the project life cycle, include tools and techniques to apply the PM skills and capabilities.
  - Product-oriented processes: Specify and create the product, defined by project life cycle, vary by application area and project phase. Required to define the scope but not defined in *PMBOK*.[1]
- **Program:** Defined as "a group of related projects, subprograms, and program activities managed in a coordinated way to obtain benefits not available from managing them individually." May include work outside the scope of projects within. A program will always have projects.
- **Program Management:** Harmonizing projects and program components, controlling interdependencies to achieve benefits outlined in the value proposition. Program Management is the application of tools/techniques and skills/knowledge to meet program requirements, to obtain benefits and control not available from managing them individually.
  - Focuses on project interdependencies.
  - Resolving resource constraints/conflicts among projects.
  - Aligning strategic direction to impact policies, procedures, processes, goals, and objectives.
  - Resolving issues and change management within governance structure.

- **Program Manager:** A project manager qualified and appointed to manage multiple concurrent projects. Program managers are typically responsible for organizing and managing the projects under a unifying program to best manage constrained resources across multiple projects. (See also Program Management.)
- **Project:** Defined as "a temporary endeavor undertaken to create a unique product, service, or result."[1] No never-ending projects. Projects must have a defined beginning and end. Projects can create a product, service, improvement, or result, e.g., a plan or document.
- **Project and Portfolio Management (PPM) Systems:** PPM software enables centralized management of processes, methods, and technologies by project managers and Project Management Offices (PMOs) to concurrently analyze and manage all proposed and active projects.
- **Project Champion:** An organizational leader with a demonstrated stake in the implementation and sustainability of a portfolio, program, project, or product. Champions will generally exert their influence to remove barriers to success and break down resistance to change. Ideally they will align with other leaders to ensure enrollment and adoption throughout the organization.
- **Project Life Cycle:** "The series of phases a project passes through from initiation to closure." Starting, organizing and preparing, project work, closing; cost and staffing levels low at the start and end; risk and uncertainty greatest at the start; ability to influence highest at start; later changes cost more.
- **Project Management:** The application of tools/techniques and skills/knowledge to meet project requirements. Project Management develops plans to achieve goals within scope, within the portfolio's goals. PMI suggests there are 47 unique project management processes within five Process Groups:
  - Initiating, planning, executing, monitoring/controlling, and closing.
  - ID requirements, address stakeholder needs, communications, managing stakeholders and creating deliverables, and balancing constraints:
    - Scope, quality, resources, budget, schedule, risks
    - Changes in one factor impact the other, e.g., scope change impact on budget, schedule
    - Changing project requirements, objectives, or schedule may create risks

- **Project Management Office (PMO):** "Management structure that standardizes the project-related governance processes and facilitates the sharing of resources, methodologies, tools, and techniques."
  - Supportive (templates, project repository); controlling (compliance); directive (managing).
  - Liaison between process/policy/procedure (PPP) and corporate measurement systems—balanced scorecard.
  - Managing shared resources across all projects within the PMO.
  - PM methodology, best practices.
  - Training and oversight.
  - Monitoring compliance via project audits.
  - Policies, procedures, and templates.
  - Communication across projects.
  - Major scope changes.
  - Optimizes shared organizational resources across all projects.
  - Manage methodologies, standards, overall risks/opportunities, metrics, and interdependencies among enterprise projects.
- **Project Management Plan (PMP):** This is iterative due to the potential for change. Progressively elaborated throughout the project life cycle, improving and detailing the plan as information and estimates become available, defining and managing work at a detailed level as the project progresses.
- **Project Management Processes:** Ensure flow throughout the project life cycle, include tools and techniques to apply the PM skills and capabilities.
- **Project Management Process Group:** "A logical grouping of Project Management inputs, tools and techniques, and outputs. PM Process Groups are not project phases." Project life cycle is categorized in five process groups:
  1. **Initiating:** Processes performed to define and authorize a new project or phase to start. Outcomes include scope and charter, overall outcome, stakeholders, project manager, and team.
     a. Business case assessment, approval, and funding are external to project boundaries, although the team may be involved in the Business Case development.
     b. Project Boundary: Point where "the start or completion of the project or phase is authorized."
     c. Key purpose of Initiation is to align stakeholder expectations with project purpose and enroll them in the scope and

objectives to illustrate how they can impact the success of the project.

2. **Planning:** Processes required to scope, refine objectives, define actions to meet objectives.
3. **Executing:** Processes performed to complete the tasks in the project plan to meet project objectives.
4. **Monitoring and Controlling:** Ongoing "background" processes required to track, review, and regulate the project progress and success in all five process stages. ID changes, evaluate, scope, and initiate changes.
5. **Closing:** Processes performed to finalize all activities and formally close out the project (or phase).

- **Project Manager:** An organizational employee, representative, or consultant appointed to prepare project plans and organize the resources required to complete a project, prior to, during, and upon closure of the project life cycle. (See also Project Management, Project Life Cycle.)
- **Project Phase:** "A collection of logically related project activities that culminate in the completion of one or more deliverables." Unique to a portion of the project or a major deliverable. Most or all process groups may be executed in each. Completed sequentially, but may overlap. Phase characteristics include:
  - Distinct focus of work from other phases; organizations, location, or skill sets may differ.
  - The Phase objective requires unique controls or processes for that phase.
  - All five Process Groups may be executed to provide added control and boundaries.
  - Phase closure includes transfer or hand-off of the deliverable, i.e., tollgate, milestone, phase review. Requires approval in most cases.
  - May be sequential, overlapping, or predictive (fully plan-driven with overlapping phases).
  - Phase examples: concept development, feasibility study, planning, design, prototype, build, test.
- **Project/Program Manager and Portfolio Leader Competencies:** Includes knowledge, performance, personal effectiveness. PMs focus on specific project objectives, control assigned project resources, and manage constraints.

- **Project Success:** Completing the project within constraints of scope, time, cost, quality, resources, and risk (as approved by PMs and senior management); the last baseline approved.
- **Project Team:** Includes PM, staff, team members. Roles include subject-matter experts, user/customer representatives, sellers, business partners.
- **Project Team Manager:** An organizational leader with a demonstrated stake in the implementation and sustainability of a portfolio, program, project, or product. The Project Team manager will generally exert his/her influence to remove barriers to success and break down resistance to change. Separate from the project manager, this is the individual who gives technical and political direction to the individuals working on a specific project/program. Ideally he/she will align with other leaders to ensure enrollment and adoption throughout the organization, and ensure his or her direct reports are working in support of the portfolio.
- **Software as a Service (SaaS):** A licensing and delivery model where software is licensed, centrally hosted, and made available as a service, typically on a subscription basis.
- **Stakeholder:** "An individual, group, or organization that may affect, be affected by, or perceive itself to be affected by a decision, activity, or outcome of a project." The project team and any "interested entities" internal or external: sponsor, customers and users, sellers, business partners, organizational groups, functional managers, other (financial, government, SMEs, consultants).
- **Strategic Business Plan:** This plan focuses on what the organization is going to do to grow its market share. It is designed to answer the questions: What do we do? How can we beat the competition? It is directed at the product?
- **Strategy:** It defines the way the mission will be accomplished. Using a well-defined strategy provides management with a thought pattern that helps it better utilize equipment and direct resources toward achieving specific goals. (For example, "The company will identify new customer markets within the United States and concentrate on expanding markets in the Pacific Rim countries.")
- **Sustaining Sponsor:** The individual/group that has the political, logistical, and economic proximity to the individuals affected by a new project/activity.

- **System:** the organizational structure, responsibilities, procedures, and resources needed to conduct a major function within an organization or to support a common business need. It is a group of processes that are required to complete a task that may or may not be connected.
- **Value:** The basic beliefs or principles upon which the organization is founded and that make up its organizational culture. They are prepared by top management and are rarely changed because they must be statements that the stakeholders hold and depend on as being sacred to the organization.
- **Value Proposition:** A document based on a review and an analysis of the benefits, costs, and value that an organization or an individual project or program can deliver to its internal/external customers, prospective customers, and other constituent groups within and outside the organization. It is also a positioning of value, where Value = Benefits − Cost (where cost includes risk).
- **Vision Statement:** Provides a view of the future desired state or condition of an organization. (A vision should stretch the organization to become the best that it can be.) The vision statement provides an effective tool to help develop objectives.

REFERENCE

1. Project Management Institute. 2013. *A guide to the Project Management Body of Knowledge (PMBOK Guide)*, 5th ed. Newtown Square, PA: PMI.

Appendix B: Potential Project and OPM/PPM Resources

CA PPM (formerly CA Clarity PPM) and CA PPM as a Service

CA Technologies' Clarity PPM (Project Portfolio Management) provides integrated Portfolio Management methodologies for different programs, projects, and portfolios while providing a holistic results-based overview of organizational performance.

According to Forrester Research, CA Technologies' "adaptable integration strategy permits companies to ... analyze performance and health to determine a portfolio alignment. Recent investments in reporting and application Portfolio Management enable CA to provide greater transparency and insight" into business planning and technology strategy alignment.

Features

- Linked investments align to key strategic initiatives transparently
- Centralized projects, resources, staffing, and budgets inform impact of change
- Constraint-based scenario comparisons help achieve best results
- Mobile time management drives better adoption and more accurate outcomes
- Create budgets and update forecasts at the summary or detail level, including labor, expenses, materials, and equipment categories

Benefits

- Make smarter portfolio decisions above-the-line
- Assure projects deliver desired results in line with market needs and business strategies
- Manage all financial aspects of your portfolio with accountability
- Gain better visibility and control of projects to speed delivery
- Drive better project execution with your choice of development methodology

In addition, organizations are seeing benefits from moving away from large, complex (often customized) implementations that require substantial overhead, upgrade, and maintenance costs. CA PPM On Demand is intended to help customers maintain a consistent release schedule while reducing overhead with a world-class Software as a Service (SaaS) solution. (From http://www.ca.com/us/intellicenter/ca-ppm.aspx)

Daptiv

Project Portfolio Management (PPM) solutions help organizations to manage demand and manage portfolios (choose the right work), manage projects and resources (do the work), and monitor performance (track the work). Daptiv PPM helps organizations in any business scenario to manage work:

- **Daptiv PPM for IT:** Provides a comprehensive set of tools to allow firms to manage their IT portfolio while maintaining alignment between IT initiatives and business priorities at all times.
- **Daptiv PPM for Enterprise PMO:** Helps companies with Enterprise Program Management Offices (EPMO) align strategically and holistically manage work across the enterprise.
- **Daptiv PPM for PMO:** Helps companies with Program Management Offices (EPMO) manage the schedule, budget, resources, and risks over numerous projects.
- **Daptiv PPM for New Product Development and Introduction (NPD/NPDI):** Helps companies with NPD or NPDI processes to shorten time to market, increase innovation, reduce new product development costs, and increase innovation.
- **Daptiv PPM for Small and Medium-Sized Business (SMB):** Helps smaller organizations manage projects and team collaboration for greater business success.
- **PPM for any business function or organization:** Can help every organization to streamline employee collaboration, reduce the complexity of managing teams, projects, and tasks, get visibility into work in the organization, and get better business results. (From http://www.daptiv.com/)

HP Project and Portfolio Management (PPM) Center

Comprehensive IT Project and Portfolio Management (PPM) software to help IT make the most of its resources and prioritize IT investments so they align with business goals and consistently deliver projects on time. Key features include:

- **Consolidated Demand Management:** Provides an overview of business needs prioritized against goals and resources by consolidating all demand. Streamline approvals and evaluate the cost and value of services.
- **Strategic Portfolio Management:** Better business decisions making with detailed, top-down planning and ad hoc queries to align IT with strategic business goals and eliminate unnecessary projects.
- **Real-Time Financial Management:** To stay informed and in charge of budgets and resources with real-time financial information on all projects. Adapt rapidly to changes in business objectives or resources.
- **Program and Project Status Dashboards:** To increase project success rates with clear dashboards on program and project status. Bring in more projects on time, scope, and budget, and reduce risk of failure.
- **Enterprise Resource Management:** Maximizes efficiency with multidimensional resource planning and tracking to help ensure best use of resources and highest-value return for the business.
- **Application Portfolio Management:** Optimizes application portfolios based on business goals. Determine which applications are of most value now and eliminate application redundancy. (From http://www8.hp.com/us/en/software-solutions/project-portfolio-management-it-portfolio-management/)

Microsoft™

Microsoft Office® Enterprise Project Management (EPM) Solution

The Microsoft Office Enterprise Project Management (EPM) Solution is an end-to-end collaborative project and portfolio environment. The Office EPM Solution helps your organization gain visibility, insight, and control across all work, enhancing decision making, improving alignment with business strategy, maximizing resource utilization, and measuring and helping to increase operational efficiency. (From http://www.microsoft.com/project/en-us/project-portfolio-management.aspx)

Microsoft Project Online: Cloud-Based Project Portfolio Management Solution

Microsoft Project Online is a flexible online solution for Project Portfolio Management (PPM) and everyday work. Delivered through Office 365, Project Online enables organizations to get started, prioritize project portfolio investments, and deliver the intended business value—from virtually anywhere on nearly any device. (From http://office.microsoft.com/en-us/project/project-portfolio-management-ppm-project-online-FX103802026.aspx)

Oracle Primavera: Professional Project Management (P6) Software Suite

Primavera P6 Professional Project Management software is designed to handle large-scale, highly sophisticated, and multifaceted projects. It can be used to organize projects up to 100,000 activities, and it provides unlimited resources and an unlimited number of target plans.

- Balance resource capacity
- Plan, schedule, and control complex projects
- Allocate best resources and track progress
- Monitor and visualize project performance
- Conduct what-if analysis and analyze alternative project plans (From http://www.oracle.com/us/products/applications/primavera/p6-professional-project-management/overview/index.html)

PD-Trak

PD-Trak is a complete Project Portfolio Management system, providing best practice templates for project selection, project execution, and support. PD-Trak offers an extensive range of tools to help you manage your project portfolio. Key objectives of project portfolio management include:

- To maximize the value of the portfolio
- To achieve balance
- To align projects with company strategy

PD-Trak provides tools aimed at (a) developing a future project portfolio plan and (b) executing the plan and monitoring progress/performance.

(From http://www.pd-trak.com/mgmttools.htm http://www.pd-trak.com/index.html)

Planview Enterprise

Planview Enterprise is an end-to-end portfolio and resource management solution designed to integrate the planning and execution of portfolios. Planview Enterprise enables you to

- capture demand (market, internal, and customer);
- prioritize portfolios (projects, applications, products, and services);
- optimize organizational capacity (people, financial, and assets);
- link plans to execution (project and resource management); and
- manage end-to-end financials (plans and actuals).

The result is a centralized platform that establishes enterprise-wide visibility into the use of resources against demands:

- Maximize opportunity with constrained resources
- Build an integrated strategy and execution plan
- Improve agility and collaboration
- Increase the capacity to innovate (From https://www.planview.com/products/planview-enterprise/)

IBM Rational Focal Point

IBM® Rational® Focal Point provides product and portfolio management driven by market needs and business objectives. This comprehensive solution helps you prioritize and select the right investments, balance change with business demands, and align resources to deliver the right products at the right time. Rational Focal Point does the following:

- **Improves decision making** by enabling you to automatically incorporate stakeholder feedback, share centralized information, and use objective information to support decisions.
- **Uses visualization, prioritization, road maps, and plans** to help you assess the effects of decisions. Now you can create plans that are achievable, value-based, and balanced against internal constraints and resources.

- **Uses predefined configurations based on best practices** to define a portfolio of investments driven by customer and business value, marketplace analysis, and stakeholder collaboration.
- **Integrates enterprise architecture and project execution into portfolio management**. This helps ensure enterprise and project decisions are aligned with your company's financial and market needs. (From http://www-03.ibm.com/software/products/en/ratifocapoin)

In an analysis of the top-ranked Project/Program Portfolio Management (PPM) Systems, the vendors were ranked from 0 to 5 (0–weak/5–strong) on elements within their current offering, ability to drive strategy, and market presence (Figure B1.1). From this analysis, CA Technologies, HP PPM Center, Daptiv and Planview solidified their position as market leaders.

SYSTEM FEATURES 1-Low (or not an area of product focus) 2-Below average; 3-Average; 4-Above avg. 5-Superior quality	CA Clarity PPM	Daptiv	GenSight	HP PPM Center	MS Office EPM	Planisware	Planview	FEATURE AVERAGE SCORE
Strategic Product Focus	4	3	1	5	3	2	3	3.0
Project Resource Management	5	3.5	4	5	4.5	5	4.5	4.5
Project Management	4.5	4	2	4	5	4.5	5	4.1
Program Implementation Support	5	5	2	5	4.5	5	5	4.5
Organizational Portfolio Management (OPM)	4	4	5	3	3	4	5	4.0
Customization and Integration Options	5	4	3	3	2	3	5	3.6
Customer Product Support and Services	2	2	1	1.5	2	1	1	1.5
Customer demand for services management	4	4	4	4	4	4	5	4.1
BI (Business Intelligence) Dashboards	4	4.5	5	4	4.5	4.5	5	4.5
OVERALL PRODUCT SCORE	4.2	3.8	3.0	3.8	3.6	3.7	4.3	3.8

Figure B1.1

Forrester Wave™: Above-the-Line Project/Program Portfolio Management, Q4 2012.

Appendix C: Sample Project Authorization Form (PAR)

Project Authorization Form (PAR)			Date : December 31, 2004	
[Your Name] [Address] [City, ST ZIP Code] [Phone] [E-mail address]		To:	PAR # [1501] PMO Attn: [Street Address] [City, ST ZIP Code] [email:]	
Project Info	**Response (Required)**		**Approver**	**Approved (Y/N)**
Portfolio or Program	Strategic Planning Program (pid: 1499)			
Project Name	Strategic Planning Process Project			
Intention	Create process (es) to align new project requests with Strategic Organizational objectives; Map the Strategic Planning Process, Project Request and Project Approval Processes			
Description	Insert detailed Project Description			
Scope/Boundaries	Processes for assembly of Strategic Plans. The actual Strategic Plan creation is not within scope for this project.			
Target beneficiary	All Departments			
Planned start date	January 1, 2015			
Planned completion date	April 1, 2015			
Budget	$10,000			
Strategic Objective alignment	Aligns with the Strategic Plan 2013–2014 Objective that all core processes impacting organizational effectiveness be mapped with the goal process optimization.			
Required support	Project sponsor will ensure resource support will be provided from one Project Manager (PM) leading a team of 4 Green Belts.			
Additional consideration	This initiative will have a beneficial ongoing impact on the organization's ability to achieve its strategic objectives.			

Index

A

abilities, *see* Skills and abilities
accountability, change management,
    122–124, *see also* Responsibilities
activities
    assigning leaders and teams, 9, 21–29
    budgets, 62–68
    business case classification, 9–10,
        29–41
    executive approval, milestones and
        budget, 11, 68–74
    identification, sponsors and managers,
        10, 56–62
    milestones, 10–11, 62–68
    prioritization, 10, 41–49
    selection, projects and programs, 10,
        49–56
activities summaries
    change management, 14, 121
    implementation plan, 12, 80–81
    OPM system implementation, 12,
        100–101
    organizational portfolio development,
        9–11, 21
    project change management
        applications, 13–14, 121
    summary, 133–134
add-ons, 18
adoption, 104, 124
agenda, meetings, 72–73
alignment, *see also* Enrollment and
    management plan
    future shock, 48
    horizontal leadership, 104
    matrix, 96
    Rational Focal Point, 152
    resources, 47, 95
    sample plan, 95
    strategic objectives, 92
Analyze phase, 65–66

appendices, sample implementation plan,
    97
application portfolio management, 149
artifacts, 90
assigning leaders and teams
    activities, 22, 27–28
    basic concepts, 21–22
    inputs, 22–27
    outputs, 22, 29
assumptions, sample plan, 92
audit example, 86
authorities, 78, 84, *see also* Project
    Management Office;
    Responsibilities
average rate of return, 35

B

background context, 51
benefit-cost ratio, 35, *see also* Costs
benefits, CA PPM, 147
best practices, 132, 152
"Bottom Zone" group, 39
budgets, *see also* Costs
    activities, 62, 63–68, 71–74
    defining, 62–68
    executive approval, 70–74
    inputs, 62, 63, 68, 70
    outputs, 62, 68, 70, 74
    overruns, 18
    personal experience, 30
    prioritization, 41
    requirements, 44, 45
    sample implementation plan, 92,
        95–97
bureaucracy *vs.* results, 134
business case
    defined, 2, 16, 135
    lack of development stage, 30–31
    milestones and budgets, 63
    validation, 31

business case classification
    activities, 9–10, 30–39
    inputs, 29, 30–31
    outputs, 9–10, 30, 39–41
Business Case Development, 30
business objectives, 20, 136
business plan, 20, 136
business processes identification, 104–105

C

CA Clarity PPM, 147–148, 153
candidate interviewing, 27–28
capacity building, 127–128
career path, 87
caution status, 107
certification, *see* Education/certification
challenges, 126–127
change, resistance to, 93, 104
change agenda
    communication, 102–104
    defined, 136
    living document, 122
    overview, 118
    parameters, 124
change agent, 8
Change Control Board, 93–94, 136
change management
    accountability, 122–124
    activities summary, 14, 121
    assessment, 123–124
    basic concepts, 13, 121
    capacity building, 127–128
    challenges, 126–127
    Change Control Board, 93–94
    communication, 126–127
    consequences, 126–127
    and control, 93–94
    integration, risk with planning,
        128–130
    management, 82, 83
    model, 122
    overview, 8, 13–14
    portfolio enrollment and management
        plan, 124–125
    project planning, 128–130
    rewards, 126–127
    risks, 126–130
    sample implementation plan, 91–94

standardization of approach, 131–132
summary, 133
sustained results, planning for, 130
Change Management Excellence: The Art
    of Excelling in Change Management,
    66
characteristics, *see* Credentials and
    characteristics
chartering, 84–85
class action lawsuit, 115–116
classification, business case
    activities, 9–10, 30–39
    inputs, 29, 30–31
    outputs, 9–10, 30, 39–41
classification models
    basic concepts, 33
    blended approach, 33, 37–39
    qualitative criteria, 33, 34
    quantitative criteria, 33, 34–37
C-level executive's pet project, 35
closing, 132, 136
Closing Process Groups, 112
cloud-based project portfolio management
    solution, 150
commitment, senior management, 92
common sense when ranking, 39
communication
    change agenda, 102–104
    change management, 126–127
    management, 81, 82
    scheduling training sessions, 103
Communication Plan, 126
Comparative Matrix, 38, 39
comparisons, software features, 153
competitive necessity category, 35
computer cost requirements, 68
concurrent dependencies, 64
configurations, predefined, 152
consensus, 101, 104, 123
consequences, change management,
    126–127
consolidated demand management, 149
constrained weighted models, 37
constraints, 92, 93
consultants, 68
content value, 47
control, change management, 93–94
Control phase, 65–66
costs, *see also* Budgets

benefit-cost ratio, 35
cost-benefit analysis, 96
estimating, 67–68
management, 81, 82
unresolved problems, *xi*
CPM, *see* Critical Path Method
credentials and characteristics, *see also*
  Job descriptions
  Portfolio Development leader/team,
    24–25, 26
  Portfolio leader, 54
criterion score, prioritization, 79
Critical Path Method (CPM), 86, 104, 105
critical success factors (CSF)
  basis for measures, 129–130
  defined, 20, 136
  OPM system implementation, 100
CSF, *see* Critical success factors
cycle time reduction, *xi*

D

Daptiv, 148, 153
dashboards, 107, 149
Data and Coordination Center, 50
data integrity procedures, 114–116
DCF, *see* Discounted cash flow
DDC Portfolio leader, 51–53
decentralization, 85–86, 104
decision making, 151
"Defer" group, 39, 43
Define phase, 65–66
deliverables, 95–97
dependencies, 92
"Desirable" group, 39
development, organizational portfolio
  activities summary, 9–11, 21
  assigning individuals and teams, 21–29
  basic concepts, 8–9, 15–16
  business case classification, 29–41
  defined, 2
  development cycle, 6–14
  executive approval, milestones and
    budget, 68–74
  flowchart, 6, 7
  identification, sponsors and managers,
    56–62
  milestones and budgets, defining,
    62–68

overview, 6–7
prioritization, 41–49
selection, projects and programs,
  49–56
typical scenario, 16–19
Development Engineering department, 17
discounted cash flow (DCF), 35
distributed scenarios, 83–84
DMAIC phases, 65–66
documentation
  business case classification, 31–32
  formal, 74
  meetings, 73
  performance requirements, 58
  resource requirements, 44, 58
  sample implementation plan, 94–95
"Do-It-Now!"" group, 38, 39, 43
dropped projects, 8–9

E

education/certification, *see also* Training
  DDC Portfolio leader, 53
  Program manager, 61
end user training, 113–114
enrollment and management plan, *see also*
  Alignment
  change agenda, 102–104
  portfolios, 124–125
  sample plan, 95
enterprise PMO, 148
enterprise resource management, 149
environmental constraints, 93
EPM, *see* Microsoft Office Enterprise
  Project Management Solution
equipment, 46, 68
Ernst & Young, 13
Executing Process Groups, 111, 143
Executive Committee, 106–110, 112,
  122–123
executive level
  activities, 11, 68–74
  approval, 11, 68, 71–74
  dashboard, 107
  inputs, 68, 70
  outputs, 11, 70, 74
  pet project, 35
  sample implementation plan, 89
  summaries, 68, 69

Executive team
  mission statement, 23–24
  presentations to, 48
experience/requirements
  DDC Portfolio leader, 53
  Program manager, 60
experts, 104, 105

F

facilities
  prioritization, 41
  requirements, 44, 45–46
failure
  reasons for, *xi*
  stopping a project before, 14
Failure Modes Effects Analysis (FMEA),
  129
Finance and Accounting department, 18
financial requirements, *see* Budgets
FMEA, *see* Failure Modes Effects Analysis
Forrester Research, 131
Forrester Ware, 153
functional competence, 60
functional groups, 104, *see also specific
  group*
future shock
  alignment factors, 48
  milestones and budgets, 66

G

Gantt chart, 71
gap analysis, 79
Gemsight, 153
glossary, sample implementation plan,
  90
goals
  accomplishing stated, 122
  basis for measures, 130
  defined, 20, 138
  going through motions, 127
good judgment, ranking, 39
governance, 78
*Guide to the Project Management Body
  of Knowledge (PMBOK Guide),* 14,
  85, 153
"gut feeling," 39

H

habits, 132
Harmful/Useful (HU) diagram, 129
Harrington, H. James, *xv–xix*
HP Project and Portfolio Management
  Center, 148–149, 153
Human Resource Department, *see also*
  Resources
  assigning leaders, 28, 50
  change agenda, 102
  job descriptions, 59

I

IBM Rational Focal Point, 151–152
identification, business processes
  impacted, 104–105
identification, sponsors and managers
  activities, 10, 56, 59, 61–62
  inputs, 56, 58
  outputs, 10, 57, 62
IDP, *see* individual development plans
impacted individuals, 8
implementation, OPM system
  activities summary, 12, 100–101
  basic concepts, 12, 99–100, 117–118
  change agenda communication,
    102–104
  critical success factors, 100
  data integrity procedures, 114–116
  end user training, 113–114
  impacted business processes
    identification, 104–105
  information technology support,
    105–107
  measurement standards, 109–110
  progress reporting, 117
  reporting, 109–110, 117
  risk identification, 110–113
  security procedures, 114–116
  standards, 109–110
  strategic goals and objectives, 101–102
  summary, 117–118
  technology constraints, 110–113
  training, 107–109, 113–114
  user acceptance training, 113–114

implementation plan
  activities summary, 12, 80–81
  assembling the team, 86–88
  basic concepts, 11–12, 77–80
  creating the plan, 88–97
  decentralization example, 85–86
  overview, 7
  PMO establishment, 83–86
  prerequisites, 81–83
  sample plan, 89–97
Improve phase, 65–66
individual development plans (IDPs), 126
Industrial Engineering department, 17
information management, 82, 83
Information Technology (IT)
  Daptiv, 148
  organizational portfolio development,
    18
  support for, 105–107
infrastructure, *see* Organizational
  Portfolio Management (OPM)
initiating and planning, *see* Planning
Initiating Process Groups, 111, 142
initiating sponsors
  defined, 8, 21, 137
  identification, 58
  milestones and budgets, 63
innovation, 2, 12
inputs
  assigning leaders and teams, 9, 22–27
  budgets, 62, 63
  business case classification, 29, 30–31
  executive approval, milestones and
    budget, 68, 70
  identification, sponsors and managers,
    56, 58
  milestones, 62, 63
  prioritization, 42, 43–44
  selection, projects and programs, 49,
    50, 54–55
integration
  Rational Focal Point, 152
  risk with planning, 128–130
internal rate of return (IRR), 35, *see also*
  Return on investment
interviewing candidates, 27–28
introduction, sample implementation
  plan, 89–90

intuition, 39
IRR, *see* Internal rate of return

J

job descriptions, *see also* Credentials and
  characteristics
  DCC Portfolio leader, 51–53
  project manager, 60–61
judgment when ranking, 39

L

large-sized businesses, 22
leadership training, 108, *see also specific*
  type of leader
"Less Desirable" group, 39
life cycle
  adaptive, 135
  predictive, 140
  project, 4
  projects, 141
  risks and constraints responsibilities,
    112
"Lower Middle Zone" group, 39
"Low-Hanging Fruit" group, 38, 43

M

Making the Case for Change: Using
  Effective Business Cases to Minimize
  Project and Innovation Failures, 16
managers, 2, 109, *see also specific type of*
  manager
manpower, 43, 67–68, *see also* People;
  Resources
Marketing and Sales department, 17
Maximizing Value Propositions to Increase
  Project Success Rates, 17
meaningful employment, 44
measurement standards, 109–110, 131
Measure phase, 65–66
medium-sized businesses
  assigning leaders and teams, 21–22
  Daptiv, 148
  prioritization, 42
meetings, 72–73

mentoring, 87
methodology, sample plan, 94
metrics, 109–110
Microsoft
    Microsoft Office Enterprise Project
        Management (EPM) Solution, 149,
        153
    Microsoft Project Online, 150
    Oracle Primavera, 150
    resources, 149–150
milestones
    activities, 10–11, 62, 63–68, 71–74
    defining, 10–11, 62–68
    executive approval, 68–74
    inputs, 62, 63, 68, 70
    outputs, 10–11, 62, 68, 70, 74
    sample implementation plan, 89–90
    time-bound projects, 128
mission statement
    assigning leaders and teams, 22–24
    business cases, 32–33
    defined, 19, 137
    Portfolio Development leader/team,
        24
    Project Management Office, 85
money, prioritization, 41
Monitoring and Controlling Process
    Groups, 111–112, 143
Monte Carlo simulations, 104–105, 129
motions, going through, 127
"must comply" criteria, 32–33
"Must Do" group, 39, 43
mutually exclusive dependencies, 64

N

"Need to Do" group, 39, 43
net present value (PV), 96–97
new product development and
    introduction (NDP/NDPI), 148
new product/service line extensions, 35
"Nice to Do" group, 39, 43
Nova Southeastern University (NSU), 85
NPD/NPDI, *see* New product development
    and introduction
NSU, *see* Nova Southeastern University

O

Office of Innovation and Information
    Technology (OIT), 85
OIT, *see* Office of Innovation and
    Information Technology
operating necessity category, 35
OPM, *see* Organizational Portfolio
    Management
Oracle software
    Primavera, 150
    Professional Project and Portfolio
        Management application, 86
organizational, 2
Organizational Change Management, 13
organizational master plan, 20, 138
organizational planning, *see* Planning
organizational portfolio development
    activities summary, 9–11, 21
    assigning individuals and teams, 21–29
    basic concepts, 8–9, 15–16
    business case classification, 29–41
    defined, 2
    development cycle, 6–14
    executive approval, milestones and
        budget, 68–74
    flowchart, 6, 7
    identification, sponsors and managers,
        56–62
    milestones and budgets, defining,
        62–68
    overview, 6–7
    prioritization, 41–49
    selection, projects and programs,
        49–56
    typical scenario, 16–19
Organizational Portfolio Management
    (OPM)
    defined, 3, 138
    development cycle, 6–14
    key element, 2
Organizational Portfolio Management
    (OPM) system
    implementation activities, 12
    implementation plan creation, 11–12
    project change management
        applications, 13–14
    structure terms, 8

Organizational Portfolio Management
   (OPM) system implementation
   activities summary, 12, 100–101
   basic concepts, 12, 99–100, 117–118
   change agenda communication, 102–104
   critical success factors, 100
   data integrity procedures, 114–116
   end user training, 113–114
   impacted business processes
      identification, 104–105
   information technology support,
      105–107
   measurement standards, 109–110
   progress reporting, 117
   reporting, 109–110, 117
   risk identification, 110–113
   security procedures, 114–116
   standards, 109–110
   strategic goals and objectives, 101–102
   summary, 117–118
   technology constraints, 110–113
   training, 107–109, 113–114
   user acceptance training, 113–114
Organizational Portfolio manager, 3
organizational process assets, 138
organizations (systematic arrangements of
   people), 3, 139
outputs
   assigning leaders and teams, 9, 22, 29
   budgets, 62, 68
   business case classification, 9–10, 30,
      39–41
   executive approval, milestones and
      budget, 11, 70, 74
   identification, sponsors and managers,
      10, 57, 62
   milestones, 62, 68
   milestones, defining, 10–11
   prioritization, 10, 42–43, 49
   selection, projects and programs, 10,
      50, 56
ownership, 26, 27

P

PAR, *see* Project Authorization form (PAR)
patterns, predictable, 127

payback period, 34
PD-Trak, 150–151
people, 41, *see also* Resources
performance requirements
   documentation, 58
pet projects, 35
Phase I, *see* Organizational portfolio
   development
Phase II, *see* Implementation plan
Phase III, *see* Organizational
   Portfolio Management system
   implementation
Phase IV, *see* Change management
Planisware, 153
planning
   implementation plan creation, 11–12
   initiating and planning, 137
   organizational, 138
   Rational Focal Point, 151
Planning Process Groups, 111, 143
Planview Enterprise, 151, 153
PMBOK, *see Guide to the Project
   Management Body of Knowledge*
PMI, *see* Project Management Institute
PMO, *see* Program Management Office;
   Project Management Office
PMP, *see* Project Management Plan
POD, *see* Portfolio Operations
   Deployment team
policy, 19–20, 139
Portfolio Development leader/team
   activities, 9, 22, 27–28
   business case validation, 31
   consensus, 123
   credentials and characteristics, 24–25,
      26
   defined, 139
   enrollment and management plan,
      124
   inputs, 22–27
   mission statement, 24
   outputs, 9, 22, 29
   presentations, 48
   purpose of, 24
   role and responsibilities, 25–26
   traits, 24–25, 26–27

Portfolio leader
  competencies, 143
  defined, 3, 5, 139
  job description, 51–53
Portfolio Operations Deployment (POD)
  team, 79
portfolios
  components, 139
  defined, 15, 139
Portfolio Steering Committee, 3, 87, 118,
  see also Steering team
position summary, 51, *see also* Job
  descriptions
PPM, 148, *see also* HP Project and
  Portfolio Management Center
prerequisites, implementation plan,
  81–83
presentations, 48, 72–73
present value, 96–97
Primavera software, 150
prioritization
  activities, 10, 42, 44–48
  basic concepts, 41–42
  implementation plan creation, 78–79
  inputs, 42, 43–44
  matrix, 80
  outputs, 10, 42–43, 49
  Rational Focal Point, 151
processes
  defined, 3, 140
  project management, 142
procurement management, 82, 83
Product Engineering department, 17
Production Control department, 17
product line extension, 35
profitability index, 35
profit/profitability models, 34–35
Program Management Office (PMO), 102
Program Management Systems, 4
Program managers
  competencies, 143
  defined, 141
  education/certification, 61
  experience, 60
  functional competence, 60
  identification, 56–62
  skills and abilities, 60–61
programmer cost requirements, 67

programs
  defined, 3–4, 140
  executive approval, 68–74
  management, 4, 140
Program Sponsors, 56–62
progress reporting, 117
Project and Portfolio Management
  systems and center, 141, 148–149
Project Authorization form (PAR), 155
Project Champion, 141
*Project Change Management: Applying
  Change Management to Improve
  Projects,* 13
Project Management Institute (PMI), 13
Project Management Office (PMO)
  chartering, 84–85
  Daptiv, 148
  defined, 4, 14
  establishment, 83–86
  mission statement, 85
  vision statement, 85
Project Management Organization, 90
Project Management Plan (PMP), 142
Project Management Process Groups,
  142–143
Project Management Systems, 4
Project managers and team
  competencies, 143
  cost requirements, 67
  defined, 4–5, 143–144
  identification, 56–62
  job description, 60–61
  milestones and budgets, 63
  participation, 109
  training, 108
projects
  components, sample plan, 90–91
  defined, 4, 141
  dropped, 8–9
  executive approval, 68–74
  failure, *xi,* 14
  importance of new, 1
  integration management, 81, 82
  life cycle, 4, 141
  management, 4, 91–94, 141–142
  phase, 143
  planning, change management,
    128–130